T0205518

# Lecture Notes in Electrical Engineering

## Volume 830

The book series *Lecture Notes in Electrical Engineering* (LNEE) publishes the latest developments in Electrical Engineering - quickly, informally and in high quality. While original research reported in proceedings and monographs has traditionally formed the core of LNEE, we also encourage authors to submit books devoted to supporting student education and professional training in the various fields and applications areas of electrical engineering. The series cover classical and emerging topics concerning:

- Communication Engineering, Information Theory and Networks
- Electronics Engineering and Microelectronics
- Signal, Image and Speech Processing
- Wireless and Mobile Communication
- Circuits and Systems
- Energy Systems, Power Electronics and Electrical Machines
- Electro-optical Engineering
- Instrumentation Engineering
- Avionics Engineering
- Control Systems
- Internet-of-Things and Cybersecurity
- Biomedical Devices, MEMS and NEMS

For general information about this book series, comments or suggestions, please contact leontina.dicecco@springer.com.

To submit a proposal or request further information, please contact the Publishing Editor in your country:

**China**

Jasmine Dou, Editor (jasmine.dou@springer.com)

**India, Japan, Rest of Asia**

Swati Meherishi, Editorial Director (Swati.Meherishi@springer.com)

**Southeast Asia, Australia, New Zealand**

Ramesh Nath Premnath, Editor (ramesh.premnath@springernature.com)

**USA, Canada:**

Michael Luby, Senior Editor (michael.luby@springer.com)

**All other Countries:**

Leontina Di Cecco, Senior Editor (leontina.dicecco@springer.com)

**\*\* This series is indexed by EI Compendex and Scopus databases. \*\***

More information about this series at https://link.springer.com/bookseries/7818

Nima Dokoohaki · Shatha Jaradat ·
Humberto Jesús Corona Pampín · Reza Shirvany
Editors

# Recommender Systems in Fashion and Retail

Proceedings of the Third Workshop at
the Recommender Systems Conference
(2021)

 Springer

*Editors*
Nima Dokoohaki
KTH - Royal Institute of Technology
Stockholm, Sweden

Shatha Jaradat
KTH - Royal Institute of Technology
Stockholm, Sweden

Humberto Jesús Corona Pampín
Spotify
Amsterdam, The Netherlands

Reza Shirvany
Digital Experience-AI and Builder Platform
Zalando SE
Berlin, Germany

ISSN 1876-1100          ISSN 1876-1119 (electronic)
Lecture Notes in Electrical Engineering
ISBN 978-3-030-94018-8          ISBN 978-3-030-94016-4 (eBook)
https://doi.org/10.1007/978-3-030-94016-4

This Springer imprint is published by the registered company Springer Nature Switzerland AG
The registered company address is: Gewerbestrasse 11, 6330 Cham, Switzerland

# Contents

# Graph Recommendations

# Using Relational Graph Convolutional Networks to Assign Fashion Communities to Users

Amar Budhiraja, Mohak Sukhwani, Manasvi Aggarwal, Shirish Shevade,
Girish Sathyanarayana, and Ravindra Babu Tallamraju

**Abstract** Community detection is a well-studied problem in machine learning and
recommendation systems literature. In this paper, we study a novel variant of this
problem where we assign predefined fashion communities to users in an Ecommerce ecosystem for downstream tasks. We model our problem as a link prediction
task in knowledge graphs with multiple types of edges and multiple types of nodes
depicting the intricate Ecommerce ecosystems. We employ Relational Graph Convolutional Networks (R-GCN) on top of this knowledge graph to determine whether
a user should be assigned to a given community or not. We conduct empirical experiments on two real-world datasets from a leading fashion retailer. Our experiments
demonstrate that the proposed graph-based approach performs significantly better
than the non-graph-based baseline, indicating that higher order methods like GCN
can improve the task of community assignment for fashion and Ecommerce users.

## 1 Introduction

The goal of community detection [13, 16, 18, 25] is to map an entity, such as a
user in an ecosystem or a node in a graph, to one or more latent communities. These
communities are subsequently used for varied tasks like fraud detection [9, 19, 21]
or personalized recommendations [10, 11, 20]. Usually, these latent communities
are discovered during the training process, and one community is different from

A. Budhiraja · M. Sukhwani (✉) · M. Aggarwal · G. Sathyanarayana · R. B. Tallamraju
Myntra Designs Pvt. Ltd., Bengaluru, India
e-mail: mohak.sukhwani@myntra.com

A. Budhiraja
e-mail: amar.budhiraja@myntra.com

G. Sathyanarayana
e-mail: girish.sathyanarayana@myntra.com

S. Shevade
Indian Institute of Science, Bengaluru, India
e-mail: shirish@iisc.ac.in

© The Author(s), under exclusive license to Springer Nature Switzerland AG 2022
N. Dokoohaki et al. (eds.), *Recommender Systems in Fashion and Retail*, Lecture Notes
in Electrical Engineering 830, https://doi.org/10.1007/978-3-030-94016-4_1

3

another in one or more latent aspects. Clustering techniques have been successfully applied in the past to perform community detection [5, 17, 23, 24]. A recent survey [17] offers an in-depth review and analysis of multiple clustering approaches. The authors broadly compare and classify multiple graph clustering approaches into multiple categories such as naive graph transformation-based, directionality preserving approaches, and methods that extend objective functions to directed networks. The iterative search algorithm for fast community detection [5], simultaneously evaluates the local importance of a node in a community and its importance concentration over all other communities to detect community members. Community detection by fuzzy clustering [23], measure similarity between nodes based on edge centrality and use fuzzy transitive rules to reveal community structures in complex graph structures. The authors partition the graph into several disjoint communities with multi-resolution thresholds. Spectral clustering-based approaches [24] have also been used in the past to identify community clusters in sparse data. The authors have described multiple approaches of encoding sparse geo-social information in graphs and identify membership communities.

Link Prediction-oriented community detection has also been studied in the existing literature [8, 14, 28] by predicting undiscovered edges in the complex networks. Cheng et al. [8] proposed two indices to compute the similarity between each pair of nodes—*A index* estimates the probability of missing links, and the *D index* recognizes spurious edges. Central node based link prediction [14] also aids community detection in complex network graphs. Local central nodes are the key nodes of communities in complex and ambiguous community structures. A joint optimization framework to learn the probability of unseen links and the probability of community affiliation of nodes simultaneously has shown promising results for real-world complex and incomplete graphs [28]. While existing tasks predict missing links and detect communities simultaneously, recent methods have mutually reinforced both to generate better results.

Inspired by the methods of link prediction for community detection, and the recent success of graph neural networks [1–4, 6, 7, 15, 27], in this paper, we propose a method to do community assignment of predefined fashion-based communities to users in an Ecommerce ecosystem. Our problem setup is motivated by real-world problems where the businesses require well-defined and explainable community definitions to understand and target potential customers for several marketing activities such as paid promotions and various discounts. We propose to model the complex Ecommerce setting through a knowledge graph with different edges for different interactions between users, products, and communities. On top of this knowledge graph, we use Relational Graph Convolutional Networks (R-GCN) [22] to perform link prediction between users and communities. We also consider a non-graph-based baseline inspired by Neural Collaborative Filtering (NCF) [12] to understand if the graph-based notion of R-GCN is helping the prediction task or not. Our empirical experiments on two real-world datasets demonstrate that the proposed R-GCN-based framework achieves notable performance and is significantly better than the non-graph-based NCF baseline.

The rest of the paper is structured as follows. In Sect. 2, we explain the proposed method followed by empirical experiments in Sect. 3. We conclude our work in Sect. 4 along with directions for future work.

## 2   Proposed Method

In this section, we discuss the proposed method. We first discuss the definitions of key terms and the details of the proposed framework for community assignment.

### 2.1   Definitions

**Definition 1** (*User*) A user is an entity in the system which performs various actions on products such as viewing them or purchasing them.

**Definition 2** (*Community*) A community is defined as a collection of items that are of the same type, from the same brand, and in the same price bucket. An example of a community could be $Tshirts\_Nike\_500 - 1000$, where the type of the product is Tshirts of Nike brand with the price range 500–1000 of relevant currency. Some more examples of product communities are: $Watches\_Fossil\_5000 - 6000$, $Trousers\_Levis\_2000 - 3000$ and $Lipstick\_Lakme\_1000 - 2000$. User implicitly becomes a member of a community if he ever purchases from the same community.

The advantage of using this definition of a community is that these are explainable in nature, unlike typical community detection algorithms, which broadly group users together using a latent notion of community. The given definition of community also gives the flexibility to use the detected user-community edges for multiple tasks such as brand-based marketing (if a user likes brand X, he or she is most likely to like brand Y), or price level uplift (will the user buy a product of a given brand from a higher price distribution).

### 2.2   Method

In order to perform community assignments to users, we define the setting in terms of a knowledge graph. The motivation for choosing a knowledge graph naturally allows us to encode the complex Ecommerce setting by representing different types of interactions through different types of edges among several entities. The knowledge graph contains three types of nodes in the proposed framework: Users, Products, and Communities. Among these three types of nodes, there are three types of edges:

**Fig. 1** Knowledge Graph
(KG) Representation of the
considered ecosystem. The
KG contains three types of
nodes—Users, Items, and
Communities and three types
of edges—View, Bought, and
Belongs To

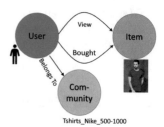

Tshirts_Nike_500-1000

(1). User-Viewed-Product (which signifies that the User viewed the Product), (2).
User-Bought-Product (which signifies that the User purchased the Product), and
User-Member-Of-Community (signifies that a User is part of a community). While
the first two edges are organic, the third edge type is constructed if the user has
ever purchased even a single item from the community. For example, if a user, $\mathcal{U}$
purchased a Nike Tshirt priced at 800 Rupees, there will be an edge between $\mathcal{U}$ and
a community called $Tshirts\_Nike\_500 - 1000$. Figure 1 shows a visualization of
the knowledge graph.

To learn the user preferences between products and communities, we employ the
R-GCN [22] on top of the knowledge graph. R-GCN is an extension of the Graph
Convolutional Network (GCN) [15]. The key formulation in GCN is to learn a feature
vector for a node in the graph based on the node's features and its neighbors (j) using
the idea of message passing between nodes as stated in the equation,

$$h_i^{l+1} = \sigma(\Sigma_{m_i \in \mathcal{M}_i} g_m(h_i^l, h_j^l)), \qquad (1)$$

where $h_i^l \in \mathbb{R}^{d^l}$ is the hidden state representation of a node $n_i$ in the $l$th layer of size
$d^l$. $\sigma$ is an element wise activation function, and $g_m$ signifies the incoming messages
and the summation of top of $g_m$ operates on the aggregation. $\mathcal{M}_i$ denotes the number
of incoming messages which often identical to number of neighbors for a given layer.

The R-GCN [22] authors augmented the model to knowledge graphs by considering message aggregation over each type of relation separately, leading to the
following formulation,

$$h_i^{l+1} = \sigma(\Sigma_{r \in \mathcal{R}} \Sigma_{m_i \in \mathcal{M}_i^r} g_m(h_i^l, h_j^l)), \qquad (2)$$

where $r$ signifies a relation, and $\mathcal{M}_i^r$ signifies the messages coming from a particular
relation. Both in GCN and R-GCN, $g_m$ is usually defined as a linear transformation
such as $Wh_j$ with a learnable weight matrix $W$. Replacing $g_m$ with $Wh_j$ leads to the
following formulation of R-GCN,

$$h_i^{l+1} = \sigma(\Sigma_{r \in \mathcal{R}} \Sigma_{j \in \mathcal{N}_i^r} (W_r^l h_j^l + W_0^l h_i^l)), \qquad (3)$$

where $N_i^r$ are the neighbors of node $n_i$ in the $r$th relation, $W_r^l h_j^l$ is a learnable weight
for aggregation of neighbors representations and $W_0^l$ is the learnable weight of self

connection of $n_i$. This formulation acts as an encoder for the link prediction task, and the authors proposed to use the DistMult method [26] as the decoder. When applied to the proposed knowledge graph, the model is expected to learn which users will purchase which product and which user belongs to which community by leveraging the extensive signals of purchasing, viewing and past community membership.

## 2.3  Model Training

As stated in the original R-GCN formulation [22], R-GCN can overfit easily owing to separate weights for each relationship in the knowledge graph. Hence, they propose the method of involving basis functions to act as a regularization R-GCN model. The basic idea is to share weights for each relation in the knowledge graph but use different coefficients on top of it to learn relation specific signals. This reduces the overall parameters of the model and helps in the regularization of the weights to prevent overfitting. The authors use the number of basis weight matrices as a hyper-parameter, which can be tuned empirically. For the link prediction formulation, as stated in the previous subsection, we optimize the cross-entropy loss where an observable triple is considered as a positive and a non-observable tuple is considered as a negative. We randomly sample five negatives per edge during training.

## 2.4  Neural Collaborative Filtering Baseline

In order to understand how significant the gains of the knowledge graph-based framework are, we define a baseline similar to Neural Collaborative Filtering Method [12]. We randomly initialize the embeddings of users, products, and communities. The relevance score between a user and product, or user and community pair, is the dot product between the embeddings of the two entities. We also add a non-linearity in the network to give it more predictive power by introducing a leaky-Relu on top of the dot product and considering the activation of the leaky-Relu as the final score between the user and product/community. We learn the parameters of this model by considering observed edges as positives and unobserved edges as negatives similar to the proposed R-GCN framework and optimize it using cross-entropy loss, similar to the R-GCN framework.

## 3  Experiments

In this section, we will discuss the empirical experiments to understand the performance of the proposed method.

## 3.1  Datasets

To perform empirical analysis, we evaluate the proposed model on two real-world datasets (one corresponding to Men's Tshirts and one corresponding to Women's Dresses) from a leading online fashion retailer in India. We consider all users who bought at least one item from the above group for four months. Table 1 gives a summary of the collected datasets in terms of how many users, communities, and products were considered. We then divide the dataset into train and test splits such that for each user, we consider half of their communities in the test split and the remaining half in train split. If the user belongs to an odd number of communities, the train split will consist of an extra community for the user than the test split. Since each product strictly belongs to only one community, we do not want the model to memorize item-to-community mapping while learning the user preferences. Therefore, we also move all the edges of the bought products belonging to communities in the test split in the test dataset. We keep all the viewed edges still as part of the training dataset because even viewed to bought is an important inference signal to learn for the end-business to identify for effective marketing and promotion strategies. The experiments empirically show how this leads to a positive lift in performance for both datasets. Table 2 gives a summary of the number of observed edges in train and test. While training, we consider five random negatives similar to the R-GCN setup for each edge. For the test time, we create a fixed set of negative edges such that these edges are neither part of the train split nor the test split. The number of negatives per user in the test set is fixed to ten (five bought edges and five belongs to edges).

**Table 1**  Dataset Stats

|            | Men Tshirts | Women tops |
|------------|-------------|------------|
| #Users     | 115646      | 53158      |
| #Communities | 473       | 318        |
| #Products  | 71923       | 49980      |
| #Edges     | 4063512     | 2519875    |

**Table 2**  Number of positive (observed) edges in train and test datasets

|            | Train dataset | | Test dataset | |
|------------|-------------|------------|-------------|------------|
|            | Men Tshirts | Women tops | Men Tshirts | Women tops |
| Belongs to | 151346      | 71516      | 174561      | 81753      |
| Bought     | 230939      | 112232     | 219095      | 92380      |
| Viewed     | 3287571     | 2161994    | NA          | NA         |

## 3.2 Hyperparameters and Best Configurations

For the R-GCN-based proposed framework, the hyperparameters include embedding size, number of basis functions, number of GCN layers, and learning rate. We set the learning rate to $10^{-3}$. For embedding size, we ran a sweep over the following sizes: $\{2^5, 2^6, 2^7, 2^8\}$; we set a number of basis functions as 1 and 2, and similarly, we evaluate GCN layers between 1 and 2. For the Neural Collaborative Filtering Baseline, we tune on learning rate and embedding size. After some hand-tuning, we set the learning rate to $10^{-3}$, and tune the embedding size over the following set: $\{2^5, 2^6, 2^7, 2^8\}$.

For the proposed framework on the Tshirts datasets, we observed the best performance on learning rate = $10^{-3}$, embedding size = $2^7$, number of basis functions = 2, and number of GCN layers = 2. For the Women Tops dataset, the best configuration is learning rate = $10^{-3}$, embedding size = $2^7$, number of basis functions to be 2, and number of GCN layers to be 1. For the Neural Collaborative Filtering baseline, the best configuration for Men Tshirts is learning rate = $10^{-3}$, and embedding size = $2^6$, and for Women Top is learning rate = $10^{-3}$, and embedding size = $2^5$.

## 3.3 Results

**Performance Metrics:** Our primary metric is means of Hits@K per user, which is the number of positives retrieved in top-K items sorted by a given score, divided by minimum of K or number of positives of the respective user. A positive indicates an existing edge in the test dataset, including both Bought or Belongs To edge types.

**Comparison of Proposed Framework with Neural Collaborative Filtering Baseline**: We compare the proposed knowledge graph-based R-GCN method with the Neural Collaborative Filtering Baseline (discussed earlier in the paper) to understand if the graph-based formulation and R-GCN is giving any gains on top of such baseline or not. Table 3 summarizes the results. The first row shows the results for Neural Collaborative Filtering Baseline, and the third row shows the results for the R-GCN-based proposed method. It can be seen that for Men Tshirts, the Hits@5 results are approximately 4% better than the baseline, and for Hits@10, the increase is higher than 1%. For Women Tops, we can see that the gap is even more significant, and the proposed framework shows more than 9% improvement for Hits@5 and more than 6% improvement for Hits@10. The results are in line with our expectations since R-GCN based framework aggregates the signals from a node's neighbors to compute the embeddings and also backpropagate the same using the labels, whereas the Neural Collaborative Filtering baseline only uses labels to tune the weights and does not take into account the neighborhood of a node.

**Comparison within Proposed Framework (With/Without View Edges)**: We also compare the version of the proposed framework where we do not consider the User-Viewed-Product edges. In Table 3, the second row shows the results without

**Table 3** Comparison with the Neural Collaborative Filtering (NCF) Baseline

| | Men Tshirts | | Women tops | |
|---|---|---|---|---|
| | Hits@5 | Hits@10 | Hits@5 | Hits@10 |
| NCF - Baseline | 0.7978 | 0.9569 | 0.7041 | 0.8769 |
| R-GCN (Without View Edges) | 0.7693 | 0.9548 | 0.7738 | 0.9372 |
| R-GCN (All Edges) | **0.8351** | **0.9682** | **0.7957** | **0.9426** |

the View Edges. Comparing it with the third row, which contains all edges, it can be seen that for Men Tshirts, the performance uplift is higher than 6% and for Women Tops, it is approximately 2% for Hits@5. For Hit@10, for both datasets, we observe a gain of 0.5% to 1% when including User-View-Edges. This indicates that View edges provide relevant signals for predicting whether a user will purchase an item and be inclined to the community.

To further understand the comparison, we also visualize the distribution of the ratio of the mean prediction probability for positive test samples to the mean prediction probability of the negative test samples per user for both the settings (with/without View Edges) in the proposed framework. We denote this metric **R**atio of mean **P**ositive score to **N**egative score (RPN). Formally, RPN for a user, $u$ is stated as

$$RPN_u = \frac{\frac{\sum_{i=1}^{n} P_i}{n}}{\frac{\sum_{j=1}^{m} P_j}{m}}, \tag{4}$$

where the user, $u$, has $n$ positive samples, $m$ negative samples, and $P_i$ is the probability of the $i$th positive sample and $P_j$ is the probability of the $j$th negative sample. If $RPN_u > 1$, it indicates that on an average positive score of a user is greater than the negative score, if $RPN_u < 1$, it indicates that negatives points have a higher probability compared to positive points for that user on average. It is desirable that the value of $RPN_u$ is always greater than 1 for each user, and the higher the value, the more confident the model will be about distinguishing positive samples from the negative ones. Figure 2b shows $RPN_u$ distributions for Men Tshirts,[1] and it can be seen that the center of gravity (i.e., mean) is higher when we consider View Edges as part of the knowledge graph. For Women Tops, we observed similar results.

**Ablation Study on Hyperparameters**: To understand how different hyperparameters affect the proposed knowledge graph-based framework, we report numbers on the three hyperparameters on Men Tshirts dataset in Table 4. For each parameter (embedding size (Emb Size), number of basis functions (#BF) and number of GCN Layers (#Layers)), we fix it to one value and compute the mean and standard deviation of Hits@K across all other parameters. It can be clearly seen that Emb Size has a similar mean performance across all values with similar standard deviation.

---

[1] We remove less than 0.05 percent outliers users for better visibility of the graphs.

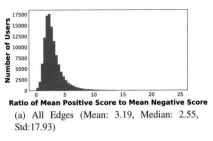
(a) All Edges (Mean: 3.19, Median: 2.55, Std:17.93)

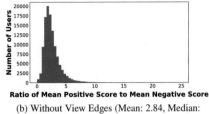
(b) Without View Edges (Mean: 2.84, Median: 2.26, Std: 19.32)

**Fig. 2** Distribution of ratio of mean positive score to mean negative score per user for men Tshirts

**Table 4** Ablation study on men Tshirts. Emb size indicates embedding dimensions, #BF indicates number of basis functions in R-GCN formulation, and #Layers indicates the number of layers of GCN used in R-GCN

| Emb size | Hits@5 | #BF | Hits@5 | #Layers | Hits@5 |
|----------|-----------|-----|-------------|---------|------------|
| $2^5$ | 0.77 (0.07) | 1 | 0.80 (0.04) | 1 | 0.79(0.08) |
| $2^6$ | 0.78 (0.08) | 2 | 0.74 (0.10) | 2 | 0.76(0.08) |
| $2^7$ | 0.77 (0.11) | | | | |
| $2^8$ | 0.76 (0.08) | | | | |

This signifies that other parameters do not impact the performance much once the embedding size is fixed. However, for the number of basis functions and number of layers, we see an observable difference indicating that basis functions and number of layers are more sensitive parameters of the model compared to embedding size.

## 4  Conclusions and Future Work

In this paper, we propose a framework to assign fashion-based explainable communities to users using their past buying patterns and viewing behavior. We use Relational Graph Convolutional Networks with a Knowledge Graph formulation to model the machine learning problem. Through our experiments, we show that the knowledge graph framework performs better than the typical Neural Collaborative Filtering Baseline. We also show that adding viewing edges to the knowledge boosts performance numbers in both datasets. As part of future work, we want to make the knowledge graph richer by introducing other edges based on payment methods, wish lists, geo-location, and searching patterns of users. We also plan to investigate how to incorporate the fashion ontology into defining and assigning these fashion-based communities. We will also look into how to include attribute information of heterogeneous nodes into the framework to improve the performance. We also plan to extend the current framework to discover fashion communities on its own by using the attributes of nodes and the past transaction history of users.

# References

1. Aggarwal M, Murty M (2021) Deep learning. Mach Learn Soc Netw
2. Aggarwal M, Murty M (2021) Machine learning in social networks: embedding nodes, edges, communities, and graphs. Springer Nature
3. Aggarwal M, Murty M (2021) Node representations. In: Machine learning in social networks. Springer
4. Aggarwal M, Murty M (2021) Region and relations based multi attention network for graph classification. In: International conference on pattern recognition (ICPR). IEEE
5. Bai L, Cheng X, Liang J, Guo Y (2017) Fast graph clustering with a new description model for community detection. Inf Sci
6. Bandyopadhyay S, Aggarwal M, Murty MN (2020) Self-supervised hierarchical graph neural network for graph representation. In: International conference on big data (Big Data). IEEE
7. Bandyopadhyay S, Aggarwal M, Murty MN (2021) A deep hybrid pooling architecture for graph classification with hierarchical attention. In: Pacific-Asia conference on knowledge discovery and data mining. Springer
8. Cheng HM, Ning YZ, Yin Z, Yan C, Liu X, Zhang ZY (2018) Community detection in complex networks using link prediction. Modern Phys Lett B
9. El Ayeb S, Hemery B, Jeanne F, Cherrier E (2020) Community detection for mobile money fraud detection. In: International conference on social networks analysis, management and security. IEEE
10. Feng H, Tian J, Wang HJ, Li M (2015) Personalized recommendations based on time-weighted overlapping community detection. Inf Manag
11. Feng L, Zhao Q, Zhou C (2020) Improving performances of top-n recommendations with co-clustering method. Expert Syst Appl
12. He X, Liao L, Zhang H, Nie L, Hu X, Chua TS (2017) Neural collaborative filtering. In: International conference on World Wide Web
13. Javed MA, Younis MS, Latif S, Qadir J, Baig A (2018) Community detection in networks: a multidisciplinary review. J Netw Comput Appl
14. Jiang H, Liu Z, Liu C, Su Y, Zhang X (2020) Community detection in complex networks with an ambiguous structure using central node based link prediction. Knowl-Based Syst
15. Kipf TN, Welling M (2016) Semi-supervised classification with graph convolutional networks. arXiv:1609.02907
16. Liu F, Xue S, Wu J, Zhou C, Hu W, Paris C, Nepal S, Yang J, Yu PS (2020) Deep learning for community detection: progress, challenges and opportunities. arXiv:2005.08225
17. Malliaros FD, Vazirgiannis M (2013) Clustering and community detection in directed networks: a survey. Phys Rep (2013)
18. Papadopoulos S, Kompatsiaris Y, Vakali A, Spyridonos P (2012) Community detection in social media. Data Min Knowl Discov
19. Peng L, Lin R (2018) Fraud phone calls analysis based on label propagation community detection algorithm. In: 2018 IEEE world congress on services (SERVICES). IEEE
20. Sahebi S, Cohen WW (2011) Community-based recommendations: a solution to the cold start problem. In: Workshop on recommender systems and the social web
21. Sarma D, Alam W, Saha I, Alam MN, Alam MJ, Hossain S (2020) Bank fraud detection using community detection algorithm. In: International conference on inventive research in computing applications. IEEE
22. Schlichtkrull M, Kipf TN, Bloem P, Van Den Berg R, Titov I, Welling M (2018) Modeling relational data with graph convolutional networks. In: European semantic web conference. Springer
23. Sun PG (2015) Community detection by fuzzy clustering. Phys A: Stat Mech Its Appl
24. Van Gennip Y, Hunter B, Ahn R, Elliott P, Luh K, Halvorson M, Reid S, Valasik M, Wo J, Tita GE et al (2013) Community detection using spectral clustering on sparse geosocial data. SIAM J Appl Math

25. Xie J, Kelley S, Szymanski BK (2013) Overlapping community detection in networks: The state-of-the-art and comparative study. ACM Comput Surveys
26. Yang B, Yih Wt, He X, Gao J, Deng L (2014) Embedding entities and relations for learning and inference in knowledge bases. arXiv:1412.6575
27. Zhang C, Song D, Huang C, Swami A, Chawla NV (2019) Heterogeneous graph neural network. In: ACM SIGKDD international conference on knowledge discovery & data mining
28. Zhang SK, Li CT, Lin SD (2020) A joint optimization framework for better community detection based on link prediction in social networks. Knowl Inf Syst

# Generative Recommendations

# What Users Want? WARHOL: A Generative Model for Recommendation

Jules Samaran, Ugo Tanielian, Romain Beaumont, and Flavian Vasile

**Abstract** Current recommendation approaches help online merchants predict, for each visiting user, which subset of their existing products is the most relevant. However, besides being interested in matching users with existing products, merchants are also interested in understanding their users' underlying preferences. This could indeed help them produce or acquire better matching products in the future. We argue that existing recommendation models cannot directly be used to predict the optimal combination of features that will make new products serve better the needs of the target audience. To tackle this, we turn to generative models, which allow us to learn explicitly distributions over product feature combinations both in text and visual space. We develop WARHOL, a product generation and recommendation architecture that takes as input past user shopping activity and generates relevant textual and visual descriptions of novel products. We show that WARHOL can approach the performance of state-of-the-art recommendation models, while being able to generate entirely new products that are relevant to the given user profiles.

**Keywords** Recommendation · Generative models · Fashion · Text-to-image · Text generation · Image generation

## 1 Introduction

### 1.1 The Growing Impact of Machine Learning on Quantitative Marketing

The activity of product marketing has four broad stages, also known as the 4Ps of Marketing, namely, *Product, Price, Promotion, and Place*. The *Product* stage deals with the activity of new product design and generates the specifications of the actual goods being created and explains their relation with the needs of the target

---

J. Samaran · U. Tanielian (✉) · R. Beaumont · F. Vasile
Criteo AI Lab, Paris, France
e-mail: u.tanielian@criteo.com

© The Author(s), under exclusive license to Springer Nature Switzerland AG 2022
N. Dokoohaki et al. (eds.), *Recommender Systems in Fashion and Retail*, Lecture Notes in Electrical Engineering 830, https://doi.org/10.1007/978-3-030-94016-4_2

audience. The second stage is the *Pricing* stage and refers to the process of deciding the base price for a product, and the possible discounting strategy. The third stage is the *Placement/Distribution* stage and defines the distribution channels that will allow the customers to have access to the new product. Finally, the final *Promotion* stage covers all type of advertising and communication around the marketing of the product.

Interestingly, current machine learning research directions cover very well the *Pricing, Placement, and Promotion* stages of marketing, but the *Product Innovation* stage has not yet been formalized as an optimization problem, and therefore remains outside the reach of quantitative methods. In our paper, we propose a way to formalize the product innovation task as a machine learning problem. Besides, to solve it, we define WARHOL, which is to the best of our knowledge, the first commercially-optimized product generation model.

## 1.2  Product Innovation as a Machine Learning Problem

The proposed goal is to automatically generate textual and visual descriptions of new products that *optimize* a given reward model, *conditioned* on specific user profiles, and name the associated task as *Personalized Optimal Product Generation (POPG)*. We argue that both innovation and personalization are the two key ingredients for formalizing Product Innovation as a Machine Learning Problem, since we believe that *the goal of product design is to explicit (in textual and visual form) the product features that optimize the utility function of the target audience.*

Notation

In order to explicit the task formally, we define $\mathbb{U}$ the space of existing users and $\mathbb{I}$ the space of existing items. Each *user $u \in \mathbb{U}$ has an embedded vectorial representation* which we assume for simplicity to be given. On the opposite, *each target item $i \in \mathbb{I}$ is the discrete tuple (text, image)* representing the target item.

We consider $\mathcal{D}_n$ be a dataset containing pairs of relevant matches of (user, items) $(u, i)$, $\mathcal{D}_n = (u, i)_{k=1,...,n}$. This type of dataset is characteristic of the next item prediction task, where given the shopping history of a user $u$ we want to predict what is the next item $i$ she will interact with.

Small reminder on classical Recommendation

Under a next item prediction formulation, the goal of content-based recommendation is to train a parametric model $F_\varphi$, $\varphi \in \Phi$ that maps each item content features $i$ into a representation that maximizes the likelihood of predicting correctly the next item:

$$\varphi^*_{RECO} = \arg\max_{\varphi \in \Phi} \; \mathbb{E}_{(u,i) \in \mathcal{D}_n} \; \ln \sigma \big( u \cdot F_\varphi(i) \big), \qquad (1)$$

where $\sigma$ refers to the softmax output function.

At inference time, given a trained recommendation model $F_{\varphi^*_{RECO}}$, we recommend the following item $i^*$ to the user $u$:

$$i^* = \arg\max_{i \in \mathbb{I}} u \cdot F_{\varphi^*_{RECO}}(i).$$

Because the likelihood function is based on the inner-product between $u$ and $F_\varphi(i)$, the best item to recommend to user $u$ is the closest item in the joint vectorial space. Here KNN represents an approximate K-Nearest Neighbors retrieval architecture that allows us to scale the search for the most similar item given the user query to large catalogs. For our experiments, we use the FAISS library implementation of HNSW [1]:

$$i^* = \arg\max_{i \in \mathbb{I}} \left( u \cdot F_{\varphi^*_{RECO}}(i) \right) = \mathrm{KNN}(u, F_{\varphi^*_{RECO}}(i)) \tag{2}$$

### Definition of POPG

For the task of POPG, one aims at finding the best generative model $G_\theta$ among a parametric class $\Theta$ that optimizes Eq. (3), where $R_\gamma(u, i)$ is an existing reward model that can evaluate any possible item given the logged the user profiles, and not only the items present in the training data, as in the case of logged reward $R(u, i)$ used in the aforementioned recommendation model. This approach is very close to the method proposed in [2]:

$$\theta^*_{POPG} = \arg\max_{\theta \in \Theta} \mathbb{E}_{u \in \mathcal{D}_n} R_\gamma\big(u, G_\theta(u)\big). \tag{3}$$

In our work, we propose a second formulation of the *POPG* objective that can use directly the logged feedback. We start from the observation that the recommendation model is a function that maps the product content to a space where the similarity between a pair of vectors is linearly correlated with the reward, e.g., the higher the similarity, the higher is the reward associated with the pair. *This means, that in the learnt space, if two product vectors are close, their associated reward for a user is also close.* We can then replace each training example with a proxy reconstruction objective, where we want to maximize the similarity between the true item $i$, and the generated item conditioned on $u$:

$$\theta^*_{POPG} \approx \arg\max_{\theta \in \Theta} \mathbb{E}_{(u,i) \in \mathcal{D}_n} F_{\varphi^*_{RECO}}(G_\theta(u)) \cdot F_{\varphi^*_{RECO}}(i). \tag{4}$$

Now, in order to use the POPG model for the Recommendation task, we propose the following:

1. Generate a set of items $k \sim G_{\theta^*_{POPG}}(u)$ conditionally to user representation $u$.
2. Get their representation $F_{\varphi^*_{RECO}}(k)$
3. Find the closest generated item to u: $F_{\varphi^*_{RECO}}(k^*) = KNN(u, F_{\varphi^*_{RECO}}(k))$
4. Finally, find the nearest existing product: $i^* = \mathrm{KNN}(F_{\varphi^*_{RECO}}(k^*), F_{\varphi^*_{RECO}}(i))$ and recommend $i^*$.

To translate in the above recommendation format, POPG is equivalent to considering that the new representation of the user becomes $F_{\varphi_{RECO}^*}(G_{\theta_{POPG}^*}(u))$.

Consequently, the POPG task requires the training of both $G_\theta$ and $F_\varphi$. It might then be more complicated than the single task of recommendation. : in the last one, one aims at learning distributions over the small space of existing items. On the contrary, POPG learns distributions in the space of all potential products.

The key contributions of our work are the following:

- To the best of our knowledge, we are the first to propose the *Personalized Optimal Product Generation (POPG)* task as a formalization of Product Innovation into a Machine Learning Problem.
- To this end, we propose the use of generative models for recommendation and develop WARHOL, which is a conditionally optimized generative model for product image and text. WARHOL can be both used to generate new personalized product ideas and to recommend existing items at similar performance levels with existing discriminative models.
- Finally, we provide strong empirical results on fashion data on both recommendation and generative tasks.

The structure of the paper is the following: In Sect. 2, we cover related work and its relationship with our proposed method. In Sect. 3, we present the WARHOL model. In Sect. 4, we present the experimental setup and the empirical results on a series of internal datasets. In Sect. 5, we summarize our findings and discuss possible extensions.

## 2 Related Work

### 2.1 Generative Models for New Item Creation:

There is an existing literature around artificial creativity and ML-driven content creation. These works range from music generation, such as the Hit Song Science task [3, 4], to story generation using recent advances in NLP as in the work by [5, 6]. As far as we know, the closest work to ours is the CogView model proposed recently by [7]. While both models both use a DALLE-like architecture, the main difference is that their model is not user-conditioned, nor optimized for any reward model.

### 2.2 Generative Models for Image Synthesis

In terms of image generation, our work builds upon existing state-of-the-art models for images retrieval and synthesis, namely CLIP [8] and DALL-E [9].

**The retrieval task** heavily relies on **CLIP** [8]. It is a model representing both text and images in a joint space. The authors have created a huge dataset of millions of paired images and texts found on the internet. Besides, similarly to word2vec [10], the model $F_\varphi$ is trained with the following objective:

$$\max_{\varphi \in \Phi} \sum_{(i_k, t_k) \in V_n} \log \sigma \left( F_\varphi(i_k) \cdot F_\varphi(t_k) \right) - \sum_k \sum_{l \neq k} \log \sigma \left( F_\varphi(i_k) \cdot F_\varphi(t_l) \right),$$

where $V_n$ is a mini-batch of $n$ pairs $(i, t)$ of texts and images. The main advantage of this model is that it can be used with success for retrieval and zero-shot classification.

**The generation part** builds on state-of-the-art text to image generation model DALL-e [9]. DALL-e is a 12-billion parameter version of GPT-3 [11] trained to generate images from text descriptions. It uses a large dataset of text image pairs. Regarding the training, the model first learns a discrete VAE that translates a high resolution image into a visual codebook. It is very similar to VQ-VAE [12]. Then, the model learns the joint distribution of both the text tokens and image tokens using an autoregressive model. The model tries to reconstruct each caption turned into a 256-long sequence of tokens with a tokenizer and, the sequence of visual tokens taken from the codebook.

Note that, in our experiments, we used the visual codebook and the encoder/decoder learned with the VQ-GAN training [13], which proved to be slightly better.

## 3   WARHOL: The Proposed Approach

We named the model WARHOL, as a recognition of both the DALL-E model and Andy Warhol, one of the first pop artists who transformed art into a commercial product.

### 3.1   The WARHOL Model

WARHOL extends the DALL-E model to the recommendation task, and therefore needs to support:

1. *Conditional generation:* being able to generate items conditionally to a user representation. In doing so, we are able to achieve personalized generations.
2. *Commercially driven generation:* being able to guide the generation toward items that have more appeal (e.g., are more likely to be viewed, bought).

To clarify, Table 1 summarizes the differences between DALL-E and WARHOL models.

**Table 1** A comparison of WARHOL versus DALL-E

|  | INPUT | OUTPUT | OBJECTIVE |
|---|---|---|---|
| DALL-E | Current item text | Current item image | Likelihood of the true current product image |
| WARHOL | CLIP of past user history | Next item text and image | Likelihood of the true next product text and image |

## 3.2 WARHOL Training

Since the training objective of WARHOL is next product prediction, as defined in (4), the generator can be decomposed in two successive steps:

1. Mapping the user state $u$ to the representation of the next product i using a mapping function $M$. Ideally, we have that this representation is close to the representation of the target item, that is $M_{\theta_1}(u) \approx F_{\varphi^*_{RECO}}(i)$.
2. Decoding $M_{\theta_1}(u)$ into sequences of text and visual tokens using a transformer-decoder $D_{\theta_2}$.

Consequently, the process of training WARHOL can be split as follows:

- **Step 0—Preprocessing:** We choose the pre-trained model CLIP as our $F_{\varphi^*_{RECO}}(i)$ and create content-based product vectors for all the products in our dataset.
- **Step 1—Training the decoder $D$:** We train a model to predict the discrete text and image of a given product conditionally to the respective CLIP embeddings. The decoder $D_{\theta_2}$ is trained to maximize the likelihood of the discrete product data, using the following objective:

$$\arg\min_{\theta_2} \mathbb{E}_{i \in \mathcal{D}_n} - \ln D_{\theta_2}(i | F_{\varphi^*_{RECO}}(i)). \tag{5}$$

- **Step 2—Training the mapping function $M$:** We start from the trained decoder $D^*_{\theta_2}$ obtained at Step 1 and freeze the parameters $\theta_2$. Now, instead of conditioning on the CLIP embedding of the item $i$, we condition the generation on the embedding of the user $u$. Note that this embedding $u$ can be either the CLIP embedding of the last item, or the average CLIP of the historical items. We add a single linear layer that maps the user representation $u$ into a representation close to $F_{\varphi^*_{RECO}}(i)$. We aim at solving the same negative likelihood minimization objective:

$$\arg\min_{\theta_1} \mathbb{E}_{(u,i) \in \mathcal{D}_n} - \ln D_{\theta_2}(i | M_{\theta_1}(u))$$

Besides, since the use of negative evidence has proven to be useful [14], we add a secondary loss term that forces the mapping function $M$ to keep the generation close to the positive example $i$ and far from a set of negatively sampled examples $i^-$. This second loss term is equivalent to a Bayesian Personalized Ranking (BPR) loss [15].

$$\arg\min_{\theta_1} \ \mathbb{E}_{(u,i)\in\mathcal{D}_n} \ -\ln D_{\theta_2}(i|M_{\theta_1}(u)) + \text{BPR}(\theta_1), \tag{6}$$

where $\text{BPR}(\theta_1) = \mathbb{E}_{(u,i)\in\mathcal{D}_n} \ -\ln\sigma\left(M_{\theta_1}(u) \cdot F_{\varphi_{RECO}^*}(i) - M_{\theta_1}(u) \cdot F_{\varphi_{RECO}^*}(i^-)\right).$

Interestingly, this second BPR term directly works in the CLIP space and skips the decoder. It is much faster to compute and, one can remove the likelihood minimization to directly optimize the simpler following objective:

$$\arg\min_{\theta_1} \ \mathbb{E}_{(u,i)\in\mathcal{D}_n} \text{BPR}(\theta_1). \tag{7}$$

We consequently distinguish two trainings for the mapping function $M$. First, $\text{WARHOL}_{\text{joint}}$ where $M_{\theta_1}$ is trained using (6). Second, $\text{WARHOL}_{\text{fast}}$ where $M_{\theta_1}$ is trained using (7).

# 4   Experiments

## 4.1   Performance on the Reco Task

### 4.1.1   Experimental Setup

The dataset is an internal fashion dataset containing 50k distinct products and 80k user product pairs (past user product, next user product), which we divide in 60k training / 20k testing sets. In order to evaluate Recommendation performance, we compute Precision@k, which computes the frequency with which positive target items are ranked in the top k among all potential targets. We report Precision@10, @100, and @1000.

To demonstrate the effectiveness of the proposed approaches, we compare WARHOL with the following recommendation baselines:

*BESTOF* represents the popularity baseline, where we compute the empirical popularity of the items in the training users in the test set. Note: All other models (baselines and WARHOL) merge their recommendations with a tuned popularity factor that favors more popular products in the recommended list.

*CLIP* represents a content-based recommendation baseline, where we compute the text and image CLIP embeddings of the past user product and recommend the top k most similar products in the CLIP space.

*RSVD* represents the collaborative filtering baseline, where we compute the Randomized SVD (RSVD) decomposition [16] of the PMI matrix product-to-product co-occurrence matrix, as proposed in [17], that proves a close connection with the

**Table 2** WARHOL performance on Recommendation task versus classical baselines

|  | Precision@10 (%) | Precision@100 (%) | Precision@1000 (%) |
|---|---|---|---|
| $BESTOF$ | 2.6 | 12 | 36 |
| $CLIP$ | 22.5 | 43 | 67 |
| $RSVD$ | 27 | 53 | 76 |
| WARHOL$_{fast}$, $s = 1$ | 7.2 | 23 | 54 |
| WARHOL$_{joint}$, $s = 1$ | 9.1 | 26.4 | 56 |
| WARHOL$_{joint}$, $s = 8$ | 10.5 | 28.9 | 59.8 |

Word2Vec algorithm [10]. As in the case of the previous baseline, we compute the best k recommendations by finding the closest k products to the user past product in the RSVD space.

### 4.1.2 Results

We compare the performance of the baselines with both WARHOL$_{fast}$ and WARHOL$_{joint}$.

The results in Table 2 show that, while not yet competitive with classical models for recommendation, such as $CLIP$ and $RSVD$, WARHOL is much better than the non-personalized baseline $BESTOF$. Also, the performance on the softer metric of Precision@1000 shows that the generated products lie in the correct general region of the vectorial space, and that the newly generated items have a good probability of being appreciated by the user.

In terms of the relative performance of two training approaches for WARHOL, we see that the joint training is systematically better than the fast objective, indicating that the decoding step in the loss, while slow, is effective. Finally, as expected, being able to sample multiple product generations and choosing the most relevant one to the user as the final recommendation query allows us to boost the ranking performance. We believe that for much larger sample sizes $s$, the performance will increase further, but due to time constraints, we were not able to run these experiments yet.

## 4.2 Conditional Product Generation

We now take a look at the generative qualities of our proposed model. In Table 3, we evaluate the relative differences in Perceptual Loss introduced in [18] between the three variants, and we observe that the image generation quality is directly proportional with the improvements in the recommendation performance. This is a good signal, meaning that improvements in the product generation quality have a measurable impact on recommendation relevance.

**Table 3** Comparison of different versions of WARHOL on image generation

|                              | Perceptual loss |
|------------------------------|-----------------|
| WARHOL$_{fast}$, $s = 1$     | 0.645           |
| WARHOL$_{joint}$, $s = 1$    | 0.612           |
| WARHOL$_{joint}$, $s = 8$    | **0.607**       |

(a) WARHOL's reconstruction ability us-    (b) Warhol's innovation ability using map-
ing decoder.                              per and decoder.

**Fig. 1** For every line, the first image is the true item whose CLIP embedding was fed to the WARHOL, followed by 5 generations

In terms of qualitative results, we show in Fig. 1 a small sample of generation results using a CLIP-to-image decoder D that is trained on the KAGGLE fashion dataset [19]. The quality of generated images is quite good, allowing the marketer to visualize the types of products the given user segment might be interested in. Furthermore, by mapping the query product through the mapping function M, we can innovate and generate more diverse products that, according to our recommendation optimized mapping, will be considered relevant by the same users.

## 5 Conclusions

In this paper, we have introduced the task of *Personalized Optimal Product Generation (POPG)* which is a formal definition of the *Product Innovation* stage of the 4-step Marketing process. In order to solve this task, we have proposed WARHOL, a product text and image generation model that is both *personalized* and *optimized* using a specific reward model. In the experimental section, we have studied the recommendation performance of our WARHOL model trained with a next item prediction reward model and showed that while it is not comparable with classical baselines based both content and collaborative filtering signal, it starts approaching them,

**Fig. 2** Comparison reconstruction versus innovation with the Warhol model

**Fig. 3** Comparison reconstruction versus innovation with the Warhol model

especially for softer metrics, such as Precision@1000. Finally, we show both quantitatively and qualitatively that WARHOL can generate high-quality descriptions of novel products.

We believe that our work show-cases a big opportunity for using generative models in recommendation, since this class of models can be used both for the default activity of matching users with existing items but also to introspect user preferences and inform the future product design cycle (Figs. 2, 3, 4 and 5).

As future work, we plan to extend our experiments to open fashion datasets and we are preparing to open-source the model and the associated training procedure. In terms of other areas of application, we would like to investigate other forms of reward models such as conversion probability and other categories of products, such as jewelry, home decor, and furniture. Finally, we are planning to work on improving the speed of text and image generation and bring it closer to real-time.

**Fig. 4** Comparison reconstruction versus innovation with the Warhol model

**Fig. 5** Comparison reconstruction versus innovation with the Warhol model

# 6 Out of Distribution on Fashion Gen

See Figs. 6, 7, 8 and 9.

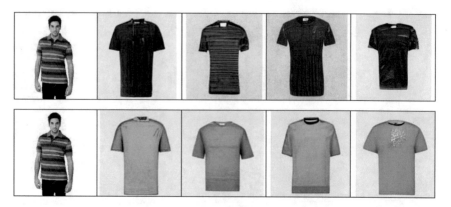

**Fig. 6** Comparison reconstruction versus innovation with the Warhol model

**Fig. 7** Comparison reconstruction versus innovation with the Warhol model

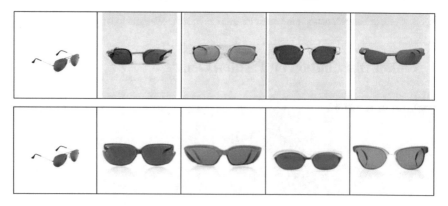

**Fig. 8** Comparison reconstruction versus innovation with the Warhol model

**Fig. 9** Comparison reconstruction versus innovation with the Warhol model

# References

1. Malkov YA, Yashunin DA (2018) Efficient and robust approximate nearest neighbor search using hierarchical navigable small world graphs. IEEE Trans Pattern Anal Mach Intell 42(4):824–836
2. Ziegler DM, Stiennon N, Wu J, Brown TB, Radford A, Amodei D, Christiano P, Irving G (2019) Fine-tuning language models from human preferences. arXiv:1909.08593
3. Ni Y, Santos-Rodriguez R, Mcvicar M, De Bie T(2011) Hit song science once again a science. In: 4th international workshop on machine learning and music. Citeseer
4. Chu H, Urtasun R, Fidler S (2016) Song from pi: a musically plausible network for pop music generation. arXiv:1611.03477
5. Fang L, Zeng T, Liu C, Bo L, Dong W, Chen C (2021) Transformer-based conditional variational autoencoder for controllable story generation. arXiv:2101.00828
6. Wang T, Wan X (2019) T-CVAE: transformer-based conditioned variational autoencoder for story completion. In: IJCAI, pp 5233–5239
7. Ding M, Yang Z, Hong W, Zheng W, Zhou C, Yin D, Lin J, Zou X, Shao Z, Yang H, Tang J et al (2021) Cogview: mastering text-to-image generation via transformers. arXiv:2105.13290
8. Radford A, Kim JW, Hallacy C, Ramesh A, Goh G, Agarwal S, Sastry G, Askell A, Mishkin P, Clark J, Krueger G et al (2021) Learning transferable visual models from natural language supervision. arXiv:2103.00020
9. Ramesh A, Pavlov M, Goh G, Gray S, Voss C, Radford A, Chen M, Sutskever I (2021) Zero-shot text-to-image generation. arXiv:2102.12092
10. Mikolov T, Chen K, Corrado G, Dean J (2013) Efficient estimation of word representations in vector space. arXiv:1301.3781
11. Brown TB, Mann B, Ryder N, Subbiah M, Kaplan J, Dhariwal P, Neelakantan A, Shyam P, Sastry G, Askell A et al (2020) Language models are few-shot learners. arXiv:2005.14165
12. Oord AV, Vinyals O, Kavukcuoglu K (2017) Neural discrete representation learning. arXiv:1711.00937
13. Esser P, Rombach R, Ommer B (2021) Taming transformers for high-resolution image synthesis. In: Proceedings of the IEEE/CVF conference on computer vision and pattern recognition, pp 12873–12883
14. Elkan C, Noto K (2008) Learning classifiers from only positive and unlabeled data. In: Proceedings of the 14th ACM SIGKDD international conference on knowledge discovery and data mining, pp 213–220
15. Rendle S, Freudenthaler C, Gantner Z, Schmidt-Thieme L (2012) BPR: Bayesian personalized ranking from implicit feedback. arXiv :1205.2618

16. Halko N, Martinsson PG, Tropp JA (2011) Finding structure with randomness: probabilistic algorithms for constructing approximate matrix decompositions. SIAM Rev 53(2):217–288
17. Levy O, Goldberg Y (2014) Neural word embedding as implicit matrix factorization. Adv Neural Inf Process Syst 27:2177–2185
18. Johnson J, Alahi A, Fei-Fei L (2016) Perceptual losses for real-time style transfer and super-resolution. In: European conference on computer vision. Springer, pp 694–711
19. Kaggle. https://www.kaggle.com/paramaggarwal/fashion-product-images-dataset

# Sizing and Fit Recommendations

# Knowing When You Don't Know in Online Fashion: An Uncertainty-Aware Size Recommendation Framework

**Hareesh Bahuleyan, Julia Lasserre, Leonidas Lefakis, and Reza Shirvany**

**Abstract**  In recent years of online fashion, the availability of large-scale datasets has fueled the success of data-driven algorithmic products for supporting customers in their journey on fashion e-commerce platforms. Very often, these datasets are collected in an implicit manner, are subjective, and do not have expert annotated labels. The use of inconsistent and noisy data to train machine learning models could potentially harm their performance and generalization capabilities. In this paper, we explore uncertainty quantification metrics within the context of online size and fit recommender systems and show how they could be used to deal with noisy instances and subjective labels. We further propose an uncertainty-aware loss function based on Monte-Carlo dropout uncertainty estimation technique. Through experiments on real data at scale within the challenging domain of size and fit recommendation, we benchmark multiple uncertainty metrics and demonstrate the effectiveness of the proposed approach for training in the presence of noise.

## 1  Introduction

E-commerce businesses have greatly benefited from creating data-driven algorithmic products to support their internal planning and to assist customers at all stages of their shopping journey. The application of machine learning models to a wide variety of tasks including product recommendations [1, 2], search and query understanding [3, 4], churn prediction [5], and price optimization [6] has enabled platforms to

H. Bahuleyan (✉) · J. Lasserre · L. Lefakis · R. Shirvany
Zalando SE, Berlin, Germany
e-mail: hareesh.pallikara.bahuleyan@zalando.de

J. Lasserre
e-mail: julia.lasserre@zalando.de

L. Lefakis
e-mail: leonidas.lefakis@zalando.de

R. Shirvany
e-mail: reza.shirvany@zalando.de

© The Author(s), under exclusive license to Springer Nature Switzerland AG 2022     33
N. Dokoohaki et al. (eds.), *Recommender Systems in Fashion and Retail*, Lecture Notes in Electrical Engineering 830, https://doi.org/10.1007/978-3-030-94016-4_3

increase customer satisfaction, business profitability, and reduce their carbon footprint [7]. One of the key driving factors that has contributed to the success of machine learning-based products in this space is the availability of large-scale customer and article datasets.

In e-commerce, and particularly in online fashion, the data used for building machine learning models is often collected in an implicit manner (e.g., user click behavior, article return data, customer purchase history, etc.) and is typically non-standardized (e.g., differences in article data information across brands). As a result, such datasets tend to be inconsistent. In the realm of size advice, subjectivity exists in multiple aspects: (i) Perspectives on fashion, style, and fit can vary greatly from one customer to another. (ii) There is no agreement with regard to sizing systems between brands (or in fact often even within a brand), and different ways exist for converting between local size systems (e.g., UK size 10 can be converted to both US size 6 and US size 8, depending on the brand). (iii) Each brand has their own target customer base, a brand might specifically target young people and use a smaller size as reference to fit garments, creating greater distortions in the larger sizes. (iv) There is "vanity sizing" [8], i.e., non-communicated and deliberate changes in physical measurements of a nominal size to boost customers' self-esteem.

Building machine learning models on top of such noisy data at scale is challenging. For instance, the generalization ability of deep neural networks has been shown to degrade when fitted with noisy data [9]. In such a situation, one intuitive albeit suboptimal solution is to discard the noisy instances and train only with the cleaner and filtered subset [10] of samples. However, a critical question still remains as to how the unreliable subset of the data can be identified.

In this work, we propose a dedicated loss based on an uncertainty estimation method to weigh instances during training. More specifically, the proposed loss function accounts for the level of uncertainty associated with each training example, which is computed using Monte-Carlo (MC) dropout uncertainty estimation [11]. Each instance's training loss is re-weighted using the standard deviation associated with its ground truth class, computed across the MC samples.

To demonstrate the effectiveness of our approach, we carry out experiments on the challenging task of personalized size recommendation where the goal is to predict the size of a customer for a given article based on their previous shopping history. The aforementioned challenges and inconsistencies in size and fit data present a strong use case to assess and demonstrate the merits of our proposal. The topic of size recommendation has increasingly become a field of focused scientific research due to its inherent challenges and its potential for game-changing impact on customer satisfaction, business profitability, and environmental issues [12–28]. Learning in the presence of noise has been well studied in the domain of image [29, 30] and text data [31, 32]. However, this topic has received only little attention within the context of recommender systems [33, 34], where the problem of large-scale dataset noise is especially relevant.

The main contributions of our work are: (1) we study and compare different uncertainty quantification metrics and their merits within the context of recommendation systems, specifically for the use case of online fashion size recommendation and (2) we formulate an uncertainty-aware loss function (based on MC-dropout uncertainty estimation) for learning in the presence of noisy data. We empirically validate its effectiveness using large-scale real data, on the challenging task of personalized size recommendation.

## 2 Related Work

Training deep neural networks in the presence of noise has been explored for computer vision [30, 35–37] and natural language processing [31, 32, 38] tasks. According to [39], the class of approaches can be broadly divided into five categories: (1) estimating noise transition matrix [40–42], (2) training with robust losses [43, 44], (3) sample selection [45, 46], (4) sample weighting [47, 48], and (5) meta learning [49, 50]. In this work, we investigate the noise issue through the data-centric approaches of sample selection and sample weighting, by integrating recent advances from the field of uncertainty estimation.

Another line of research is to incorporate abstention or rejection mechanism to the model during the training stage. This enables the model to refrain from making a prediction in a situation of high uncertainty or noise [29, 51]. However, this may require making changes to the model or training setup through auxiliary losses. Our approach, on the other hand, is simpler, only requiring pruning or re-weighting individual training instances.

The scientific literature proposes several methods for uncertainty quantification [11, 52–56]. We adopt Monte-Carlo dropout uncertainty estimation [11] due to its simplicity and computational efficiency. Within recommender systems, the uncertainty topic has been studied with respect to optimizing exploration/exploitation strategy when making new recommendations [57], assessment of human behavioral uncertainty [58], product ranking after accounting for prediction uncertainty [59], and data uncertainty in IoT-based recommender systems [60].

As highlighted in Sect. 1, in this work, we study the topics of uncertainty estimation and learning in the presence of noise from an application point of view to size recommendation. The task of personalized size recommendation has been studied using Bayesian modeling [16, 23], neural collaborative filtering [17, 61], sequential models [20, 62], and more recently using meta-learning approaches [18]. In this work, we use the transformer-based self-attention model [63] for size recommendation introduced in [20]. However, we point out that our proposed framework is general and applicable to a wide range of datasets and models which use deep neural networks.

## 3   Approach

In this section, we begin by formulating the personalized size recommendation task. Next, we provide a description of Monte-Carlo (MC) dropout which forms the basis for our uncertainty estimation. This is followed by a discussion of different metrics which can be used for uncertainty-based decision-making. We finally present the proposed uncertainty-aware training framework.

### 3.1   Size Recommendation Task Formulation

In the context of fashion e-commerce, machine learning-based personalized size recommendation systems are trained to estimate the size of a customer for a given garment based on their previous purchases. We use the notation $C$ to represent a customer whose purchase history consists of a sequence of $T$ articles $\{a_1, \ldots, a_T\}$, which forms the input for a supervised learning algorithm. This set of past purchases by the customer is referred to as *support purchases*. Note that each article is represented as a feature vector consisting of multiple fashion attributes (see Sect. 4.1). Next, when $C$ is faced with a new *query article* $a_{T+1}$, the model is required to infer the appropriate size that the article should be purchased in. The model predicts the most probable size $v$ for $a_{T+1}$, from the set $\mathcal{V}$ of possible sizes which represents the $V$-dimensional output space of class labels. In this work, since both the support purchases and the query articles are constructed from the past purchases of customers, the ground truth sizes are available. The support purchases of $C$ along with the query article and its true size are treated as a training instance for the model. This is used to train the model with categorical cross-entropy (CE) loss:

$$\mathcal{L}_{CE} = -\sum_{i=1}^{N} \log p_{\tilde{v}}^{(i)} \tag{1}$$

where $p_{\tilde{v}}^{(i)}$ is the probability assigned to the ground truth size $\tilde{v}$ by the model for training instance $i$ and $N$ is the total number of training instances.

### 3.2   Uncertainty-Based Decision-Making

Machine learning models trained with noisy data might generate predictions with high uncertainty. For e-commerce platforms, it is important to quantify this uncertainty, particularly when interacting with customers. In the size recommendation application, uncertainty information helps in deciding when to confidently show a size recommendation to the customer vs. when to take an alternative action, such as suggesting the customer to order two sizes and keep only the one that fits best. This depth of understanding helps gain and retain customer trust. In this section, we first describe an uncertainty estimation technique and then present various metrics for quantifying predictive uncertainty.

### 3.2.1 Monte-Carlo (MC) Dropout at Test Time

A popular regularization technique adopted in training deep neural networks is dropout [64]. In the work by [11], applying dropout during the *testing* phase was put forward as a simple method for estimating prediction uncertainty by generating multiple outputs for the same input, based on multiple forward passes with different sets of network weights enabled/disabled. This technique is referred to as MC dropout due to the sampling of different sets of model weights at test time. In our work, MC dropout is used to derive some of the uncertainty metrics that we describe in Sect. 3.2.2. We use the notation $M$ to represent the number of MC samples drawn for a given test instance. Let $\bar{\mathbf{p}}_m$ be the $V$-dimensional output probability *vector* corresponding to sample $m$. The mean probability vector can be obtained by taking an average across the $M$ samples, i.e., $\bar{\mathbf{p}}_{mean} = \frac{1}{M} \sum_{m=1}^{M} \bar{\mathbf{p}}_m$. Furthermore, we denote the *scalar* value of the probability assigned to class $v$ in sample $m$ as $p_{m,v}$ and in the mean vector as $p_{mean,v}$. In a similar manner, we denote $\bar{\mathbf{p}}_{std}$ to be the vector of standard deviations computed across MC samples and $p_{std,v}$ to be the scalar standard deviation associated with the $v$th class.

### 3.2.2 Metrics for Quantifying Predictive Uncertainty

Next, we outline below different metrics for quantifying the predictive uncertainty associated with the size recommended by the model for a given test instance.

1. **Maximum Softmax Score** (max_proba): The probability assigned to the predicted size.
2. **Difference between the first and second highest predicted probabilities** (diff): The difference between the probabilities assigned to the first and second most probable sizes.
3. **Predictive Entropy** (entropy): This refers to the entropy of the output softmax distribution and is calculated as $\text{entropy} = -\sum_{v \in \mathcal{V}} p_v \log(p_v)$.
4. **Bayesian Active Learning by Disagreement** (bald): Proposed by [65], this metric is closely related to predictive entropy. It is computed as $\text{bald} = -\sum_{v \in \mathcal{V}} p_{mean,v} \log(p_{mean,v}) - \frac{1}{M} \sum_{m=1}^{M} \left[ -\sum_{v \in \mathcal{V}} p_{m,v} \log(p_{m,v}) \right]$, i.e., the difference between the entropy of the averaged distribution and the average entropy over distributions.
5. **MC-Dropout Normalized Standard Deviation-Predicted Size** (stddev_pred): From $\bar{\mathbf{p}}_{std}$ obtained by MC sampling, we report as an uncertainty metric

the standard deviation associated with the predicted size $\hat{v} = \arg\max_{v \in \mathcal{V}} \bar{\mathbf{p}}_{mean,v}$ normalized by the average predicted probability of $\hat{v}$ from $\bar{\mathbf{p}}_{mean}$, i.e., $\texttt{stddev\_pred} = \frac{p_{std,\hat{v}}}{p_{mean,\hat{v}}}$.

6. **MC-Dropout Variation Ratio** (`var_ratio`): Variation ratio measures the disagreement (if any) between size predictions across the MC samples by a voting system calculated as $\texttt{var\_ratio} = 1.0 - \frac{M_{majority}}{M}$, where $M_{majority}$ is the number of votes for the most frequently predicted size.

7. **Entropy From MC-Dropout Samples** (`dropout_entropy`): With `var_ratio`, we only look at the agreement or disagreement with regard to the most frequent size among the MC samples. Introduced in [66], `dropout_entropy` additionally computes an entropy score based on the number of votes obtained by each of the predicted sizes.

For `entropy`, `bald`, `stddev_pred`, `var_ratio`, and `dropout_entropy`, higher values imply higher uncertainty. For the remaining metrics, namely, `max_proba` and `diff` a lower value corresponds to a higher uncertainty.

**MC-Dropout Normalized Standard Deviation of the True Size**

One of the metrics that we introduced earlier in this section for uncertainty quantification is `stddev_pred`, the standard deviation associated with the *predicted size*, obtained from MC-dropout samples. Our benchmarking experiments (see Sect. 5.1), however, revealed that the standard deviation associated with the *ground truth* size divided by its mean predicted probability was a stronger indicator of the model's predictive uncertainty. We refer to this metric as `stddev_true`, calculated as $\texttt{stddev\_true} = \frac{p_{std,\tilde{v}}}{p_{mean,\tilde{v}}}$, where $\tilde{v}$ refers to the ground truth size for the given test instance. It is to be noted that the use of this metric at test time is a hypothetical scenario as we do not have access to the true target label. However, we use `stddev_true` within the proposed uncertainty-aware training framework for hard and soft weighting of training instances as described in the following section.

## 3.3 Uncertainty-Aware Framework for Learning in the Presence of Noise

### 3.3.1 Hard Weighting: Clean Training Data by Pruning

Uncertainty-based decision-making, i.e., value thresholding of a metric to avoid making an erroneous prediction in an uncertain scenario is a test time strategy. However, it is important to note that models may encounter such high uncertainty instances even at training time as a result of dataset noise, thereby potentially hindering the training process. To overcome this issue, we propose to use the uncertainty metric `stddev_true` to clean up the training data as follows:

- **Stage I-Training**: We use a $k$-fold training strategy wherein we train the model $k$ times and hold out a different subset of the training data each time.
- **Stage II-Uncertainty Estimation**: Once training is complete, we compute the uncertainty metrics on each of the $k$ hold out sets using the respective instances of the model trained on the complementary subsets.
- **Stage III-Dataset Pruning**: Next, using `stddev_true` as the uncertainty score, we filter out the instances with the highest uncertainty or noise, thereby retaining only a pre-defined proportion of the original training dataset.
- **Stage IV-Re-training**: The model is once again trained from scratch using only this new, cleaned, subset.

The $k$-fold strategies in Stage I and II address to an extent the issues of overfitting and underestimation of uncertainty in contrast to a singlefold training and testing of the model on the same dataset. With our four-stage procedure, the objective is to discard the noisy instances and re-train in a way that the model captures the correct learning signals for the problem at hand. However, one question that arises is what proportion of the dataset should be pruned, if we do not know beforehand the percentage of noise in the data. Moreover, all pruned instances may not all necessarily be noise. Some may be more challenging instances that the model could have benefited from. To address this issue, we next discuss an improved procedure to deal with the noisy instances, rather than the binary hard-weighting approach of filter in or filter out.

### 3.3.2 Soft Weighting: Keep and Re-weight All Training Instances

Instead of simply pruning a subset of the training data, we can use the uncertainty metric to weight the contribution of each training instance in the re-training step. To this end, we propose an uncertainty-aware loss function, where the normalized standard deviation corresponding to the true size obtained from MC dropout (`stddev_true`) re-weights the loss contribution of each training instance. Thus, the categorical cross-entropy loss in Eq. 1 is modified as follows:

$$\mathcal{L}_{reweighted-CE} = -\sum_{i=1}^{N} \exp(-\beta \sigma^{(i)}) * \log p_{\tilde{v}}^{(i)} \tag{2}$$

where $\sigma^{(i)}$ refers to `stddev_true` associated with training instance $i$ and $\beta$ is a hyperparameter. The contribution of instances, to the loss, that have a higher degree of noise is down-weighted. During training, Stages I and II remain the same as in Sect. 3.3.1, Stage III is skipped, and Stage IV uses Eq. (2).

# 4    Experimental Setup

## 4.1    Dataset, Model, and Evaluation Metric

For both dataset and model, we follow the setup adopted by [20]. The details of the customer purchase dataset used in our experiments and the train/validation/test splits are provided in Table 1. The size recommendation model is built on a transformer-based encoder-decoder architecture. The input to the model is the sequence of customer purchases consisting of article embeddings based on product attributes (such as high-level category, gender, fashion category, brand, season, and supplier), purchase timestamps encoded as positional embeddings, and size embeddings which denote the size the article was purchased in. The configuration of the encoder and decoder blocks used is the same as in [20]. The maximum length of support purchases is fixed at 40 and the batch size is set to 256. All model variants are trained for 20k iterations using SGD with momentum and a multiplicative learning rate decay of 0.9. For MC-dropout uncertainty estimation, the dropout rate and number of test time samples are set as 0.3 and 10, respectively.

For the proposed uncertainty framework using `stddev_true`, we have two main hyperparameters—the percentage of training data pruned in the hard-weighting experiments and the $\beta$ term in the loss function for the soft-weighting experiments. We discuss below our design choices for these hyperparameters, obtained by tuning on the validation set. As for evaluation metrics, we report top-1 accuracy for size prediction at different levels of test set coverage. For instance, given an uncertainty metric and a coverage level of $p$ %, we filter out the most uncertain test instances that fall within the top $1 - p$ percentile based on value thresholding of the uncertainty metric and report the accuracy on the remaining $p$ % subset. The accuracies are reported at coverage levels of 100%, 80%, 60%, and 40%. For the hyperparameter tuning experiments, we use `max_proba` as the uncertainty thresholding criteria, motivated by the results presented in Sect. 5.1.

### 4.1.1    Percentage of Dataset Pruned

Discarding too many instances when pruning might considerably reduce the size of the training data while discarding too few might not truly address the noise issue. We experiment by varying the percentage of training instances pruned. As illustrated in Table 2, we observe empirically that pruning 20% of the data seems to provide a reasonable accuracy coverage trade-off on the validation set and this pruning rate is

**Table 1**  Dataset statistics

| #customers | #purchases | #articles | #brands | #sizes | train/val/test split |
|------------|------------|-----------|---------|--------|----------------------|
| 260k       | 5.7M       | 603k      | 2221    | 1162   | 252k/18k/172k        |

**Table 2** Pruning rate selection for hard-weighting experiments by tuning on the validation set. In each row, we report the accuracies obtained at coverage levels 100%, 80%, 60%, and 40%, when the specified percentage of training instances was pruned based on `stddev_true` metric

| Pruned percentage | 100% | 80% | 60% | 40% |
| --- | --- | --- | --- | --- |
| 10% | 61.11 | 67.21 | 71.59 | 76.72 |
| 20% | **62.21** | **68.04** | **73.39** | **78.79** |
| 30% | 62.20 | 67.87 | 73.17 | 78.12 |
| 40% | 61.73 | 67.76 | 73.17 | 77.75 |

**Table 3** Selecting $\beta$ for soft-weighting experiments by tuning on the validation set. In each row, we report the accuracies obtained at coverage levels 100%, 80%, 60%, and 40%, when $\beta$ in the loss function (Eq. 2) is varied

| $\beta$ | 100% | 80% | 60% | 40% |
| --- | --- | --- | --- | --- |
| 1.0 | 61.63 | 67.79 | 73.53 | 78.49 |
| 3.0 | 61.87 | 68.07 | **73.63** | **79.24** |
| 5.0 | 61.79 | **68.11** | 73.52 | 79.00 |
| 10.0 | **61.97** | 67.98 | 73.17 | 78.53 |

used in our test experiments. We point out that the choice of percentage pruned can vary between datasets, depending on the inherent noise level. This issue of having to determine the pruning rate can be circumvent by the instance level soft-weighting strategy which uses the entire training data.

### 4.1.2 Strength of Loss Re-weighting Penalty Factor

We analyze in Table 3 the effect of the $\beta$ hyperparameter in the re-weighted cross-entropy loss function (Eq. 2). Our experiments on the validation set shows that $\beta = 1.0$ does not sufficiently penalize noisy instances and results in a sub-optimal model. While higher values of $\beta$ provide a stronger penalizing effect on the noisy instances, the choice of $\beta$ is observed to be dependent on the desired coverage level of the application. We set $\beta = 3.0$ in our experiments.

## 5 Results

### 5.1 Comparison of Uncertainty Metrics

We base our analysis on the idea that if a metric indicates a high degree of uncertainty associated with a prediction, we may be better off to refrain from showing a recommendation or instead take an alternate action, rather than making an erroneous

**Table 4** Comparison of uncertainty metrics. Each row represents the size prediction accuracies on the test set reported at coverage levels 100%, 80%, 60%, and 40%, when the respective uncertainty metric is used for thresholding

| Uncertainty metric | 100% | 80% | 60% | 40% |
|---|---|---|---|---|
| max_proba |  | **66.92** | 72.28 | 77.89 |
| diff |  | 65.20 | 70.80 | 76.41 |
| entropy |  | 66.72 | 71.57 | 77.61 |
| bald | 60.42 | 62.73 | 65.10 | 66.43 |
| stddev_pred |  | 65.47 | 70.34 | 75.93 |
| var_ratio |  | 64.64 | 64.61 | 64.54 |
| dropout_entropy |  | 64.64 | 64.61 | 64.54 |
| stddev_true |  | **74.74** | **90.01** | **98.15** |

prediction. This corresponds to filtering out the most uncertain predictions and catering to a subset of the customer-article pairs. Table 4 shows the accuracies obtained at different test set coverage levels for the various uncertainty metrics discussed in Sect. 3.2. For instance, when thresholding with entropy at a coverage level of 80%, we filter out test instances that fall within the top 20% quantile based on entropy and compute the accuracy on the remaining 80% subset. Note that for 100% coverage level, since no thresholding is applied, all metrics have the same accuracy.

We observe that the accuracy of var_ratio and dropout_entropy remains unchanged beyond a coverage level of 80%, since beyond a certain proportion of the test set, all MC samples agree on the predicted size for a given input. However, the predicted probability vectors differ between the MC samples, which is indicated by the increasing accuracy for stddev_pred metric. Still, it is outperformed by entropy and diff and the best performance comes from max_proba. In other words, simply using the maximum softmax score as an uncertainty metric to threshold at the desired coverage is a strong baseline. This finding has also been previously reported in [67]. We note that under a hypothetical case, stddev_true outperforms max_proba. Although this metric cannot be used for thresholding since the true size is unknown at test time, we show in the next section how using it in the uncertainty-aware training framework provides performance gains.

## 5.2 Dataset Pruning and Re-weighting

We compare our approach of using stddev_true for dataset pruning and re-weighting to two categories of baselines. In the first group, we include other promising uncertainty metrics such as max_proba, entropy, and stddev_pred. Note that in contrast to other uncertainty metrics, since a lower max_proba indicates higher uncertainty, we negate its value in the soft-weighting experiment. In the second category, we compare our framework with other competitive approaches from recent literature [29, 42] which have been studied on image and text data. To the best of our knowledge, this is the first time they are being applied to a recommendation task. The model in [29], which we refer to here as Deep Abstaining Recommender, is trained to abstain from making a recommendation for instances that have a high associated uncertainty, as determined by the probability assigned to the abstain class. In [42], a method called confident learning was employed, which estimates a latent joint distribution between the given noisy labels and unknown corrupted labels. For the above-mentioned approaches, as way to prune the dataset, we follow the criteria prescribed in the respective papers: (1) the probability assigned to the abstain option for Deep Abstaining Recommender and (2) the normalized probability margin based on [68] for confident learning. We further use the same criteria to re-weight the loss function similar to Eq. 2 to compare against our proposed `stddev_true`-based re-weighting. In all cases, `max_proba` is used at test time as the thresholding criteria for obtaining accuracies at different coverage levels as it performed best in Table 4.

As shown in Table 5, the baseline which simply trains on the full noisy data (without re-training) is sub-optimal. We highlight the effectiveness of using `stddev_pred` and `stddev_true` for dataset pruning and re-weighting over `max_proba`

**Table 5** Results of hard-weighting (dataset pruning) and soft-weighting (instance re-weighting) experiments. The top block corresponds to the baseline approach of training with the entire dataset as is and using `max_proba` as the thresholding criteria at test time. The middle and the bottom blocks show the results of the dataset pruning and dataset re-weighting experiments, respectively, each applying a different metric for pruning/re-weighting. Each row represents the size prediction accuracies on the test set reported at coverage levels 100%, 80%, 60%, and 40%, when the respective method is used

| | Coverage level ⟶ | 100% | 80% | 60% | 40% |
|---|---|---|---|---|---|
| Training on full (noisy) dataset | Baseline | 60.42 | 66.92 | 72.28 | 77.89 |
| Re-training with hard-weighting | max_proba | 60.25 | 66.90 | 72.44 | 78.00 |
| | entropy | 60.21 | 66.90 | 72.45 | 78.25 |
| | stddev_pred | 60.53 | 67.26 | 72.66 | **78.74** |
| | Deep Abstaining Recommender | 60.44 | 67.01 | 72.33 | 78.19 |
| | Confident Learning | **61.80** | **67.86** | **73.09** | 78.33 |
| | stddev_true | 61.09 | 67.11 | 72.30 | 77.31 |
| Re-training with soft-weighting | max_proba | 60.36 | 66.83 | 72.21 | 78.25 |
| | entropy | 57.21 | 64.01 | 69.33 | 74.99 |
| | stddev_pred | 60.67 | 67.27 | 72.52 | 78.34 |
| | Deep Abstaining Recommender | 60.45 | 67.02 | 72.60 | 78.38 |
| | Confident Learning | **61.34** | **67.76** | 73.04 | **78.63** |
| | stddev_true | 61.16 | 67.59 | **73.04** | 78.62 |

and `entropy`. Confident learning [42] for pruning or loss re-weighting during model re-training achieves the best performance. Note, however, that, in the case of loss re-weighting, `stddev_true` performs almost on par with confident learning, although it only estimates uncertainty locally from the true class probability instead of having to compute joint distributions across multiple classes. This results in confident learning having a computational complexity which is quadratic in the number of classes [42], whereas with `stddev_true` we attain comparable performance with linear complexity. Finally, for both hard- and soft-weighting experiments, `stddev_true` performs better than Deep Abstaining Recommender.

## 6 Conclusion

We study the issues of noise and uncertainty in recommendation systems. Using fashion size recommendation as an application, we analyze and compare different uncertainty quantification metrics. Motivated by the learnings from the metric comparison, we explore two methods to address the issue of noisy data instances: (1) hard weighting via dataset pruning based on uncertainty metric thresholding and (2) soft weighting via instance re-weighting through uncertainty-aware loss. Our experiments show that using the normalized standard deviation associated with the probability of the true size, obtained via Monte-Carlo uncertainty estimation, for instance, weighting is effective for learning in the presence of noise. We recommend this soft-weighting strategy because the noise associated with each training instance is directly accounted for, rather than the hard-weighting strategy where one has to know in advance the inherent dataset noise level to decide the pruning percentage. For future work, we wish to explore ideas from semi-supervised learning such as co-training [69], which could circumvent the need for multi-stage training. Experimentation of our approach on domains such as images and text is also left for future work.

## References

1. Wu C, Yan M (2017) Session-aware information embedding for e-commerce product recommendation. In: Proceedings of the 2017 ACM on conference on information and knowledge management, pp 2379–2382
2. Hwangbo H, Kim YS, Cha KJ (2018) Recommendation system development for fashion retail e-commerce. Electron Commer Res Appl 28:94–101
3. Karmaker Santu SK, Sondhi P, Zhai C (2017) On application of learning to rank for e-commerce search. In: Proceedings of the 40th international ACM SIGIR conference on research and development in information retrieval, pp 475–484

4. Haldar M, Abdool M, Ramanathan P, Xu T, Yang S, Duan H, Zhang Q, Barrow-Williams N, Turnbull BC, Collins BM, Legrand T (2019) Applying deep learning to airbnb search. In: Proceedings of the 25th ACM SIGKDD international conference on knowledge discovery and data mining, pp 1927–1935

5. Berger P, Kompan M (2019) User modeling for churn prediction in e-commerce. IEEE Intell Syst 34(2):44–52

6. Wu L, Hu D, Hong L, Liu H (2018) Turning clicks into purchases: revenue optimization for product search in e-commerce. In: The 41st international ACM SIGIR conference on research and development in information retrieval, pp 365–374

7. Nestler A, Karessli N, Hajjar K, Weffer R, Shirvany R (2021) Sizeflags: reducing size and fit related returns in fashion e-commerce. In: Proceedings of the 27th ACM SIGKDD international conference on knowledge discovery and data mining

8. Weidner NL (2010) Vanity sizing, body image, and purchase behavior: a closer look at the effects of inaccurate garment labeling. Master's Thesis

9. Zhang C, Bengio S, Hardt M, Recht B, Vinyals O (2021) Understanding deep learning (still) requires rethinking generalization. Commun ACM 64(3):107–115

10. Frénay B, Verleysen M (2013) Classification in the presence of label noise: a survey. IEEE Trans Neural Netw Learn Syst 25(5):845–869

11. Gal Y, Ghahramani Z (2016) Dropout as a Bayesian approximation: representing model uncertainty in deep learning. In: International conference on machine learning, pp 1050–1059. PMLR

12. Walsh G, Mohring M, Koot C, Schaarschmidt M (2014) Preventive product returns management systems: a review and a model. In: Proceedings of the 22nd European conference on information systems

13. Zhang Y, Juhlin O (2015) Using crowd sourcing to solve the fitting problem in online fashion sales. Glob Fash Manag Conf 1:62–66

14. Diggins MA, Chen C, Chen J (2016) A review: customer returns in fashion retailing. Anal Model Res Fash Bus 31–48

15. Cullinane S, Browne M, Karlsson E, Wang Y (2019) Retail clothing returns: a review of key issues. Contemp Oper Logist 301–322

16. Guigourès R, Ho YK, Koriagin E, Sheikh AS, Bergmann U, Shirvany R (2018) A hierarchical bayesian model for size recommendation in fashion. In: Proceedings of the 12th ACM conference on recommender systems, pp 392–396. ACM

17. Sheikh A-S, Guigourès R, Koriagin E, Ho YK, Shirvany R, Vollgraf R, Bergmann U (2019) A deep learning system for predicting size and fit in fashion e-commerce. In: Proceedings of the 13th ACM conference on recommender systems, pp 110–118

18. Lasserre J, Sheikh AS, Koriagin E, Bergman U, Vollgraf R, Shirvany R (2020) Meta-learning for size and fit recommendation in fashion. In: Proceedings of the 2020 SIAM international conference on data mining, pp 55–63. SIAM

19. Lefakis L, Koriagin E, Lasserre J, Shirvany R (2020) Towards user-in-the-loop online fashion size recommendation with low cognitive load. Recomm Syst Fash Retail Springer 734:59

20. Hajjar K, Lasserre J, Zhao A, Shirvany R (2021) Attention gets you the right size and fit in fashion. Recommender systems in fashion and retail, Springer, pp 77–98

21. Abdulla GM, Borar S (2017) Size recommendation system for fashion e-commerce. In: KDD workshop on machine learning meets fashion

22. Sembium V, Rastogi R, Saroop A, Merugu S (2017) Recommending product sizes to customers. In: Proceedings of the eleventh ACM conference on recommender systems. ACM, pp 243–250

23. Sembium V, Rastogi R, Tekumalla L, Saroop A (2018) Bayesian models for product size recommendations. In: Proceedings of the 2018 world wide web conference, pp 679–687

24. Du ES, Liu C, Wayne DH (2019) Automated fashion size normalization. CoRR, arXiv:abs/1908.09980

25. Vecchi A, Peng F, Al-Sayegh M (2015) Looking for the perfect fit? Online fashion retail-opportunities and challenges. In: Conference proceedings: the business and management review, vol 6, pp 134–146. The Academy of Business and Retail Management

26. Karessli N, Guigourès R, Shirvany R (2019) Sizenet: weakly supervised learning of visual size and fit in fashion images. In: IEEE conference on computer vision and pattern recognition (CVPR) workshop on FFSS-USAD
27. Hsiao W-L, Grauman K (2020) Vibe: dressing for diverse body shapes. In: Computer vision and pattern recognition
28. Han X, Wu Z, Wu Z, Yu R, Davis LS (2018) Viton: an image-based virtual try-on network. In: Proceedings of the IEEE conference on computer vision and pattern recognition, pp 7543–7552
29. Thulasidasan S, Bhattacharya T, Bilmes J, Chennupati G, Mohd-Yusof J (2019) Combating label noise in deep learning using abstention. In: International conference on machine learning, pp 6234–6243. PMLR
30. Li J, Socher R, Hoi SC (2020) Dividemix: learning with noisy labels as semi-supervised learning. In: 8th international conference on learning representations, ICLR 2020, Addis Ababa, Ethiopia
31. Jindal I, Pressel D, Lester B, Nokleby M (2019) An effective label noise model for dnn text classification. In: Proceedings of the 2019 conference of the North American chapter of the association for computational linguistics: human language technologies, pp 3246–3256
32. Tayal K, Ghosh R, Kumar V (2020) Model-agnostic methods for text classification with inherent noise. In: Proceedings of the 28th international conference on computational linguistics: industry track, pp 202–213
33. Said A, Jain BJ, Narr S, Plumbaum T (2012) Users and noise: the magic barrier of recommender systems. In: International conference on user modeling, adaptation, and personalization. Springer, pp 237–248
34. Yera TR, Mota YC, Martinez L (2015) Correcting noisy ratings in collaborative recommender systems. Knowl-Based Syst 76:96–108
35. Zhang Z, Sabuncu MR (2018) Generalized cross entropy loss for training deep neural networks with noisy labels. In: 32nd conference on neural information processing systems (NeurIPS)
36. Han J, Luo P, Wang X (2019) Deep self-learning from noisy labels. In: Proceedings of the IEEE/CVF international conference on computer vision, pp 5138–5147
37. Huang J, Qu L, Jia R, Zhao B (2019) O2u-net: a simple noisy label detection approach for deep neural networks. In: Proceedings of the IEEE/CVF international conference on computer vision, pp 3326–3334
38. Van Landeghem J, Blaschko M, Anckaert B, Moens MF (2020) Predictive uncertainty for probabilistic novelty detection in text classification. In: Proceedings ICML 2020 workshop on uncertainty and robustness in deep learning
39. Cordeiro FR, Carneiro G (2020) A survey on deep learning with noisy labels: how to train your model when you cannot trust on the annotations? In: Proceedings of the 33rd SIBGRAPI conference on graphics, patterns and images (SIBGRAPI). IEEE, pp 9–16
40. Patrini G, Rozza A, Krishna Menon A, Nock R, Qu L (2017) Making deep neural networks robust to label noise: a loss correction approach. In: Proceedings of the IEEE conference on computer vision and pattern recognition, pp 1944–1952
41. Hendrycks D, Mazeika M, Wilson D, Gimpel K (2018) Using trusted data to train deep networks on labels corrupted by severe noise. Adv Neural Inf Process Syst 31:10456–10465
42. Northcutt C, Jiang L, Chuang I (2021) Confident learning: estimating uncertainty in dataset labels. J Artif Intell Res 70:1373–1411
43. Liu Y, Guo H (2020) Peer loss functions: learning from noisy labels without knowing noise rates. In: International conference on machine learning, pp 6226–6236. PMLR
44. Ma X, Huang H, Wang Y, Romano S, Erfani S, Bailey J (2020) Normalized loss functions for deep learning with noisy labels. In: International conference on machine learning, pp 6543–6553. PMLR
45. Han B, Yao Q, Yu X, Niu G, Xu M, Hu W, Tsang I, Sugiyama M (2018) Co-teaching: robust training of deep neural networks with extremely noisy labels. In: Bengio S, Wallach HM, Larochelle H, Grauman K, NicolCesa-Bianchi, Garnett R (eds) Advances in neural information processing systems 31: annual conference on neural information processing systems 2018, December 3–8, 2018, Montréal, Canada, pp 8536–8546

46. Nguyen DT, Mummadi CK, Ngo TP, Nguyen TH, Beggel L, Brox T(2020) SELF: learning to filter noisy labels with self-ensembling. In: 8th international conference on learning representations, ICLR 2020, Addis Ababa, Ethiopia
47. Wang Y, Liu W, Ma X, Bailey J, Zha H, Song L, Xia ST (2018) Iterative learning with open-set noisy labels. In: Proceedings of the IEEE conference on computer vision and pattern recognition, pp 8688–8696
48. Lee KH, He X, Zhang L, Yang L (2018) Cleannet: transfer learning for scalable image classifier training with label noise. In: Proceedings of the IEEE conference on computer vision and pattern recognition, pp 5447–5456
49. Ren M, Zeng W, Yang B, Urtasun R (2018) Learning to reweight examples for robust deep learning. In: International conference on machine learning, pp 4334–4343. PMLR
50. Shu J, Xie Q, Yi L, Zhao Q, Zhou S, Zongben X, Meng D (2019) Meta-weight-net: learning an explicit mapping for sample weighting. Adv Neural Inf Process Syst 32:1919–1930
51. Geifman Y, El-Yaniv R (2019) Selectivenet: a deep neural network with an integrated reject option. In: International conference on machine learning, pp 2151–2159. PMLR
52. Abdar M, Pourpanah F, Hussain S, Rezazadegan D, Liu L, Ghavamzadeh M, Fieguth P, Cao X, Khosravi A, Acharya UR et al (2021) A review of uncertainty quantification in deep learning: techniques, applications and challenges. Inf Fus
53. Lakshminarayanan B, Pritzel A, Blundell C (2017) Simple and scalable predictive uncertainty estimation using deep ensembles. In: Advances in neural information processing systems
54. Heo J, Lee HB, Kim S, Lee J, Kim KJ, Yang E, Hwang SJ (2018) Uncertainty-aware attention for reliable interpretation and prediction. In: Proceedings of the 32nd international conference on neural information processing systems, pp 917–926
55. Xiao Y, Wang WY (2019) Quantifying uncertainties in natural language processing tasks. Proceedings of the AAAI conference on artificial intelligence, vol 33, pp 7322–7329
56. Maddox WJ, Izmailov P, Garipov T, Vetrov DP, Wilson AG (2019) A simple baseline for bayesian uncertainty in deep learning. Adv Neural Inf Process Syst 32:13153–13164
57. Zeldes Y, Theodorakis S, Solodnik E, Rotman A, Chamiel G, Friedman D (2017) Deep density networks and uncertainty in recommender systems. In: Proceedings of TargetAd conference, London, United Kingdom
58. Jasberg K, Sizov S (2017) Assessment of prediction techniques: the impact of human uncertainty. In: International conference on web information systems engineering. Springer, pp 106–120
59. Zhang M, Guo X, Chen G (2016) Prediction uncertainty in collaborative filtering: enhancing personalized online product ranking. Decis Support Syst 83:10–21
60. Liu X, Guojun Wang Md, Bhuiyan ZA, Shan M (2020) Towards recommendation in internet of things: an uncertainty perspective. IEEE Access 8:12057–12068
61. Dogani K, Tomassetti M, Vargas S, Chamberlain BP, De Cnudde S(2019) Learning embeddings for product size recommendations. In: eCOM@ SIGIR
62. Eshel Y, Levi O, Roitman H, Nus A (2021) Presize: predicting size in e-commerce using transformers. In: Proceedings of the 44th international ACM SIGIR conference on research and development in information retrieval
63. Vaswani A, Shazeer N, Parmar N, Uszkoreit J, Jones L, Gomez AN, Kaiser L, Polosukhin I (2017) Attention is all you need. In: Advances in neural information processing systems, pp 5998–6008
64. Srivastava N, Hinton G, Krizhevsky A, Sutskever I, Salakhutdinov R (2014) Dropout: a simple way to prevent neural networks from overfitting. J Mach Learn Res 15:1929–1958
65. Houlsby N, Huszar F, Ghahramani Z, Lengyel M (2011) Bayesian active learning for classification and preference learning. CoRR, arXiv:abs/1112.5745
66. Zhang X, Chen F, Lu CT, Ramakrishnan N (2019) Mitigating uncertainty in document classification. In: Proceedings of the 2019 conference of the north american chapter of the association for computational linguistics: human language technologies, vol 1 (Long and Short Papers), pp 3126–3136

67. Geifman Y, El-Yaniv R (2017) Selective classification for deep neural networks. Adv Neural Inf Process Syst 30:4878–4887
68. Wei C, Lee J, Liu Q, Ma T (2018) On the margin theory of feedforward neural networks. CoRR, arXiv:abs/1810.05369
69. Blum A, Mitchell T (1998) Combining labeled and unlabeled data with co-training. In: Proceedings of the 11th annual conference on computational learning theory, pp 92–100

# SkillSF: In the Sizing Game, Your Size is Your Skill

**Hamid Zafar, Julia Lasserre, and Reza Shirvany**

**Abstract** Rating systems are popular in the gaming industry to maximize the uncertainty of the win/loss outcome of future games by pairing players with similar skills thanks to ranking players through a score representing their "true" skill according to their performance in past games. Inspired by this approach, we propose to tackle the challenging problem of online size recommendations in fashion by reformulating a customer's purchases as multiple sizing games where we denote (a) the return status of an article as a win/loss (size-related return) or as a draw (no size-related return) and (b) the garment/body dimensions or size as the skill. This re-framing allows us to leverage rating systems such as TrueSkill [1] and to propagate semantically meaningful information between article sizes and customer sizes. Through experimentations with real-life data, we demonstrate that our approach SkillSF is competitive with the state-of-the-art size recommendation models and offers valuable insights as to the underlying true size of customers and articles.

## 1 Introduction

Garment and shoe sizing is extremely complex. In addition to the lack of standardization across countries and categories, and to the multiple co-existing size systems, brands size their garments differently according to their target audience. With several thousands of unique values to represent sizes, picking a size without seeing the garment has become a massive challenge in online fashion and shopping platforms have needed to actively support customers through size advice. Offering size advice

H. Zafar (✉) · J. Lasserre · R. Shirvany
Zalando SE, Berlin, Germany
e-mail: hamid.zafar@zalando.de

J. Lasserre
e-mail: julia.lasserre@zalando.de

R. Shirvany
e-mail: reza.shirvany@zalando.de

not only increases customer satisfaction but can also reduce costs dramatically and save a lot of polluting returns and packaging.

Algorithmic size advice is a recent and growing academic topic. Current models are typically based on (a) return data [2], (b) purchase histories [3–10], (c) dialogues with the customer [11–13], (d) images and computer vision [11, 14–19] and (e) combinations of the first 4 points [19–21]. The focus of this work is on personalized size recommendations from customers' purchase histories (the second category). The main approaches in the first category are statistical [3, 5, 20] or deep-learning [8, 9, 22] methods. The statistical approaches usually need a conversion system to map the sizes into a numerical scale, whereas the deep-learning methods learn customers and articles embedding in an end-to-end fashion. We are interested in an approach that can benefit from the numerical scale.

A common issue faced in personalized size recommendations is that very little is known about the article other than category/brand, and even less about the customer besides their purchase history. While gathering information about articles is expensive, especially for each size they come in, asking customers about their body shape or for pictures is delicate due to privacy and to the risk of degrading their experience. Yet, we hypothesize that if a few garments have meaningful dimensions measured accurately—for example { length, width, sleeve length, neck size }—these measurements could propagate to customers and, through customers, to other garments for which we do not have measurements.

The gaming industry gives an interesting perspective on the topic. Player rating approaches such as TrueSkill [1] have been developed (a) to infer the "true" skill of players based on the observation of the outcome of their past games, and (b) to match players in future games so as to maximize the uncertainty around the outcome or the probability of a draw. Looking at the sizing problem, one can consider customers and (sized) articles to be players and purchases to be sizing games, where the players are assessed on skills related to sizing. In such games, a draw indicates that the garment and the customer are of similar sizes and a potential match, while a win/loss indicates that the purchased item is bigger/smaller than the customer and may be returned. Here the skill should be a set of dimensions shared by both types of players and predictive of whether a garment will be kept or returned for a size issue. In the remainder of this paper we refer to an article in a particular size as an SSKU (Simple Stock Keeping Unit).

In this work, we cast the problem of the size recommendations as a player and game ranking problem and introduce SkillSF, a size recommender based on ranking strategies. The contributions are threefold:

- We show through experiments on real data that this SkillSF is competitive with state-of-the-art size recommenders.
- We demonstrate that this approach allows us to infer the latent numerical sizes of customers and articles.
- In contrast to previous work, sizes are modeled at the SSKU level so the different behaviors across sizes of the same garment can be captured.

## 2  SkillSF

In this approach, each customer and each SSKU is a player and their skill is represented by a set of dimensions common to both player types, here their (true) numerical size. This true size is unobserved and treated as latent variable. Each purchase is a game between an SSKU player $a_i$ and a customer player $c_j$, which leads to updates for both customer and SSKU numerical sizes. The observations are the outcomes of these games: a win is a purchase returned because the article was too big, a draw is a kept item, a loss a purchase returned because the article was too small.

### 2.1  Skill Inference

We follow TrueSkill [1] to model the relationship between skills and game outcomes and to infer the posterior distribution of the players' latent numerical size. We model $p\left(\mathbf{r}\,|\mathbf{s}\right)$, the probability of a game outcome given the true sizes of the involved players. TrueSkill can handle multiple players and teams however in this work a team is always 1 player so we will omit the concept, and there are only 2 players at a time, so $\mathbf{r}$ can be reduced to $r = 1$ for a win by player 1, $r = 0$ for a draw and $r = -1$ for a win by player 2. Note that gaming is best served by online training so we go through one game at a time and update the distribution parameters accordingly. Applying Bayes theorem, we get the posterior distribution of the true sizes given the game just played using

$$p\left(s_{a_i}, s_{c_j}\,\middle|\,r_{a_i c_j}\right) = \frac{p\left(r_{a_i c_j}\,\middle|\,s_{a_i}, s_{c_j}\right) p\left(s_{a_i}\right) p\left(s_{c_j}\right)}{p\left(r_{a_i c_j}\right),} \tag{1}$$

where $a_i$ is the $i$th SSKU, $c_j$ the $j$th customer and $r_{a_i c_j}$ the outcome of their game. Note that the sizes of the players are independent so the prior $p\left(s_{a_i}, s_{c_j}\right)$ can be factorized. The prior $p\left(s_k\right)$ on the size of player $k$ is a Gaussian such that $p\left(s_k\right) = \mathcal{N}\left(s_k\,\middle|\,\mu_k, \sigma_k^2\right)$, as we know the actual dimensions of customers fluctuate daily according to how they eat/sleep, and the dimensions of SSKUs fluctuate across instances. After each game, the posterior becomes the prior for the next game. Given their size, each player is then expected to perform at a level $p_k$ drawn from $\mathcal{N}\left(p_k\,\middle|\,s_k, \beta^2\right)$, where $\beta$ is indicative of how far customer and SSKU sizes should be for a win/loss to be likely. Finally the game outcome is modeled using the performance: the winning outcome $r_{a_i c_j} = 1$ requires $p_{a_i} > p_{c_j} + \epsilon$ and the draw outcome $r_{a_i c_j} = 0$ requires $\left|p_{a_i} - p_{c-j}\right| < \epsilon$, where $\epsilon > 0$ is the draw margin. The likelihood is then given by

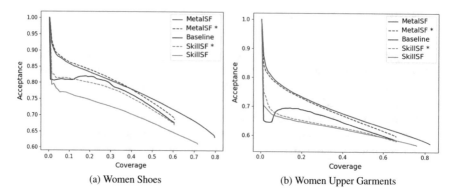

(a) Women Shoes                    (b) Women Upper Garments

**Fig. 1** Acceptance vs coverage as the minimum imposed on the difference between the top-2 probability scores changes (*Only on test purchases covered by Baseline)

$$p\left(r_{a_i c_j} \middle| s_{a_i}, s_{c_j}\right) = \int_{-\infty}^{\infty} p\left(r_{a_i c_j} \middle| p_{a_i}, p_{c_j}\right) p\left(p_{a_i} \middle| s_{a_i}\right) p\left(p_{c_j} \middle| s_{c_j}\right) dp_{a_i} dp_{c_j}, \quad (2)$$

where, for example, $p\left(r_{a_i c_j} = 1 \middle| p_{a_i}, p_{c_j}\right) = I\left(p_{a_i} > p_{c_j} + \epsilon\right)$.

The joint distribution $p(\mathbf{s}, \mathbf{p}, \mathbf{r})$ of a game is represented by the factor graph shown in Fig. 1 of [1]. *We do not consider teams so the intermediate variable $t_i$ in Fig. 1 of [1] (which stands for team performance) and the individual performance $p_i$ are always linked by the identity factor, and that we consider 2 players (2 branches) only.* The posterior distribution of the sizes is approximated by a Gaussian and inference is achieved using message passing. All the necessary equations and details for training are provided in [1] and we built on existing code from https://trueskill.org/.

## 2.2 Size Recommendation

Once the model is trained using the outcome of all historical purchases, all players have updated distributions for their true size $s$ and, based on these true sizes, the model can be used to predict the outcome of future purchases. These outcome predictions can in turn be leveraged to give a size recommendation for a customer C and a garment A by predicting the outcome of the games between C and all the SSKUs associated with A (i.e., all possible sizes of A) and choosing the SSKU with the closest probability to draw. Additionally, in our case, the true sizes are semantically meaningful and can provide insights about SSKUs and customers.

## 2.3   Hyper-parameters

All in all, the hyper-parameters of the model are:

- $\beta$ and $p_d$, which are used to calculate the draw margin $\epsilon = \sqrt{2}\beta \; \Phi^{-1}\left(\frac{p_d+1}{2}\right)$, where $\Phi$ is the cumulative distribution function of the canonical Gaussian. $\beta$ specifies the difference needed between two players' skills for an 80% chance of winning, and the model considers two players to be a match if the difference of their skills is within the draw margin $\epsilon$.
- all the initial $\{\mu_k, \sigma_k^2\}$ for the players' size priors at the beginning of training, i.e., before any game is played. By default, we initialize customers with the mean SSKU's mapped size found in the training set and a high variance of 4, while SSKUs are initialized using the mapped size and a low variance (0.5)

# 3   Experimental Set-Up

*Data*. We use purchase data from a particular market between 2016-01-01 and 2020-09-30, where the purchased items were either kept or returned due to a size-related issue (e.g., the item was too big or too small). We choose two categories: Women Shoes and Women Upper Garments. Table 1 provides the overall insights about the datasets.

Note that the distributions of the number of purchases per SSKU or customer are long tailed. About 80% of SSKUs have fewer than 4 associated purchases for Women Shoes, fewer than 7 for Women Upper Garments, while 80% of customers have fewer than 18 purchases for Women Shoes, fewer than 27 for Women Upper Garments.

*Sizes*. The sizes for Women Shoes are numerical (35–49), while the sizes for Women Upper Garments come from different size systems, mostly confection (numerical,

**Table 1**  Description of the dataset

| | # Unique values | # Customer# SSKU | Train validation test | % Purchases with known customer/SSKU/Both |
|---|---|---|---|---|
| Women shoes | 3,069,113 828,845 | 394,723 | 78%/10%/12% | 0.75/0.23/0.003 |
| Women upper garments | 7,755,774 | 407,202 1,523,221 | 1,523,22178%/11%/11% | 0.80/0.19/0.001 |
| Women upper garments (confection sizes only) | 3,651,777 | 339,997 772,067 | 79%/10%/9% | 0.79/0.18/0.0008 |

30–48) or alpha (4XS to 4XL). Handling the lack of standardization in sizes is out of scope of this work so we use expert-designed rules to convert alpha sizes to numerical confection sizes. This conversion is an approximation and does not reflect the truth for every brand.

*Other methods*. One of the first size recommenders proposed was given as a baseline in [20]. It relies on a customer size distribution which is essentially a Gaussian where the mean and standard deviation are computed using the sizes in the customer's kept purchase history. We refer to this model as Baseline. Baseline is trained on each category separately and assumes numerical sizes (therefore requires size conversion). Recent developments in size recommendation have put forward deep-learning models [8, 10]. Here we use MetalSF [9] as comparison, a meta-learning-based approach also used successfully in [21] to combine customer data and purchase data. MetalSF is trained jointly on both categories. It does not require size conversions, as sizes are treated as strings and the latent spaces provided by the inner layers of the model exploit correlations across purchases to learn such conversions through rich size representations.

*Pre-processing*. For Baseline and SkillSF, unless otherwise specified, alpha sizes are converted to numerical confection sizes. We follow Baseline and filter out purchases that have a large distance to the customer's average purchased size (higher than 1 for Shoes, than 2 for Upper Garments) to increase the quality of the data. Such cases happen when the customer purchases an article for someone else. Customers with high standard deviation in purchased sizes (higher than 2 for Shoes, than 4 for Upper Garments) are further discarded. However, none of these pre-processing steps are applied to MetalSF as to the mapped sized are not available to MelalSF [9]. Additionally, Baseline requires 2 training purchases for a customer to receive recommendations [20].

*Train/validation/test split*. We split the dataset according to time using the purchase timestamp in order to mimic the real-world setting where we can only use past purchases to predict sizes for future purchases. The training set is shuffled before each training epoch. Experiments indicate no difference when training purchases are sorted by time. Most test purchases are associated with customers or articles that have very low numbers of historical purchases, e.g., more than 80% of SSKUs have no purchases in the training set for both Women Shoes and Women Upper Garments, and about 30% of customers have no purchases in the training set. While performance is reported on all test purchases, the training set is generally restricted to kept purchases. Returned purchases are included in training in Sect. 4.5 only.

*Metrics.* We will report results on the following 4 metrics:

- *Coverage* represents the proportion of purchases for which the model was able to provide a size recommendation, regardless of its relevance.
- *Acceptance* is the proportion of given size recommendations that were followed by the customer, i.e., that matched the customer purchased size.
- *Accuracy* is the proportion of given size recommendations that were helpful to the customer given by accuracy $= \frac{TP+TN}{\text{number of given recommendations}}$, where $TP$ is the number

of accepted recommendations not followed by a size-related return, and $TN$ the number of rejected size recommendations followed by a size-related return.

- *Precision* is the proportion of accepted size recommendations that were kept given by precision $= \frac{TP}{TP+FP}$, where $FP$ the number of accepted size recommendations followed by a size-related return.

# 4 Results

## 4.1 Hyper-Parameters Tuning

We performed a grid search on the validation set to find the optimal values for $\beta$ and $p_d$ summarized in Table 2.

The resulting draw margin in both Shoes and Upper Garments on kept data is around 75% of a size step (1 for Shoes, 2 for Upper Garments), indicating no tolerance for the next size but quite a large tolerance for intra-size variations. This draw margin drops dramatically when including return data, in particular for Upper Garments, indicating much less tolerance for variations.

## 4.2 Size Recommendation

Throughout the Results section, the task is to recommend a given customer a size for a given article that is likely to fit them well, size chosen among all the sizes available for that article. For Baseline and MetalSF, all the sizes of an article (i.e., all related SSKUs) get a probability score and the top-scoring SSKU is recommended. For SkillSF, the SSKU with the highest probability of drawing in a match with the customer is recommended.

Table 3 shows the performance of Baseline, MetalSF, and SkillSF for Women Shoes and Women Upper Garments.

**Table 2** Hyper-parameters values for SkillSF

|  | $\beta$ | Draw probability $p_d$ | Draw margin $\epsilon$ |
|---|---|---|---|
| Women shoes | 0.50 | 0.70 | 0.73 |
| Women upper garments | 1.00 | 0.70 | 1.46 |
| Women upper garments initialized with confection sizes only | 1.00 | 0.70 | 1.46 |
| Women shoes with training return data | 0.5 | 0.2 | 0.18 |
| Women upper garments with training return data | 0.8 | 0.1 | 0.14 |

**Table 3** Performance metrics computed on the test set for all models

|                       |          | Coverage | Acceptance | Accuracy | Precision |
|-----------------------|----------|----------|------------|----------|-----------|
| Women shoes           | Baseline | 0.60     | 0.67       | 0.58     | 0.78      |
|                       | MetalSF* | 0.60     | 0.68       | 0.62     | 0.78      |
|                       | SkillSF* | 0.60     | 0.67       | 0.62     | 0.79      |
|                       | MetalSF  | 0.80     | 0.63       | 0.63     | 0.79      |
|                       | SkillSF  | 0.72     | 0.65       | 0.61     | 0.79      |
| Women upper garments  | Baseline | 0.66     | 0.58       | 0.51     | 0.80      |
|                       | MetalSF* | 0.66     | 0.59       | 0.57     | 0.80      |
|                       | SkillSF* | 0.66     | 0.58       | 0.56     | 0.80      |
|                       | MetalSF  | 0.82     | 0.57       | 0.60     | 0.80      |
|                       | SkillSF  | 0.76     | 0.56       | 0.55     | 0.80      |

*Only on test purchases covered by Baseline

Since these approaches vary in pre-processing steps (and since Baseline requires 2 training purchases to provide a recommendation), they support varying amounts of the test set. Hence, we also compute the metrics on the subset of the test set supported by Baseline (denoted with a star). Overall the 3 models perform very similarly on the common subset of the test set, with a better accuracy for MetalSF and SkillSF. The acceptance of SkillSF slightly degrades when allowing its full coverage, but much less than the acceptance of MetalSF due to the presence of the unfiltered noisy customers. Figure 1 shows how performance evolves when setting a minimum on the difference between the top-2 probability scores returned by the model across all possible sizes. This minimum is only used to rank the test samples: the top-scoring size is recommended, however the difference with the second best size accounts for more certainty. Moving this minimum generates different values for all the metrics, here we compute acceptance vs coverage.

We can see that SkillSF performs roughly on par with Baseline (dashed green and solid black lines). Baseline (solid black line) performs poorly on high scoring probabilities/low coverage regime. It is unclear why.

## 4.3 Size Initialization

As described in Sect. 2, the model assumes a prior $\mathcal{N}\left(s_k \mid \mu_k, \sigma_k^2\right)$ on the size of player $k$. Model training starts with initializing this prior for all players. For customers, it can be initialized (a) either with the mean SSKU size found in the training set and a high variance of 4, or (b) with the mean and variance of the sizes in their kept training purchases. We tried both in various proportions and saw no significant impact on performance.

**Table 4** Metrics computed on Women Upper Garments for both SSKU initialization strategies

| SSKU initialization | Coverage | Acceptance | Accuracy | Precision |
|---|---|---|---|---|
| (a) confection sizes only (about 50% SSKUs) | 0.74 | 0.34 | 0.40 | 0.79 |
| (b) all SSKUs through size conversion | 0.76 | 0.56 | 0.55 | 0.80 |

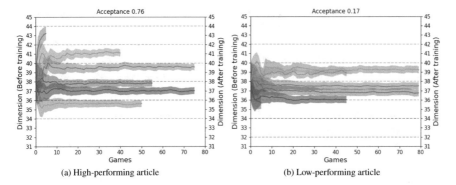

(a) High-performing article                    (b) Low-performing article

**Fig. 2** Changes in non-confection articles sizes (Women Upper Garments) when SkillSF is initialized using confection sizes only

For SSKUs, it can be initialized either (a) using the label size when numerical and a low variance (0.5) or the average label size of the SSKUs with numerical label sizes in the training set and a high variance (2), or (b) using the label size of the SSKU when numerical or the converted label size and a low variance (0.5). Table 4 shows the performance when customers are initialized with the mean SSKU size from the training set for the 2 possible SSKU initialization strategies.

While the performance of SkillSF is lower when using only the confection label sizes, Fig. 2 shows that the existing information can reach the articles with no confection sizes through customers who purchased both types of articles. In other words, SkillSF is able to learn the sizes of customers or articles from the purchases. We can imagine doing the same with the measurements such as waist, hips, or shoulders, which would allow information propagation from a few anchor customer/articles with known measurements. The dotted lines show the converted sizes as noisy ground truth. For the high-performing article on the left-hand side, the size distributions are well ordered and have little to no overlap. For the low-performing article, it is the opposite.

## 4.4 Insights into Article Sizes

This section provides further analysis into the articles' sizes from two perspective: First, we illustrate the impact of games in the training process on the development of articles' latent size distribution. Second, we investigate to what extend the learnt sizes are in line with the size and fit feedback from experts, as well as the size and fit issue raised by SizeFlags [2], a probabilistic Bayesian model that provides a article-based size recommendation.

***Skill journey***. Fig. 3 shows the development of the latent size distributions of an article during training, with a different color for each size. Both articles have about 200 purchases and 11 SSKUs. The article on the left-hand side shows strong performance with 72% acceptance; the size distributions are stable, with little to no overlap. The size distributions of the article on the right-hand side show much more overlap, consequently the recommendations on this article are poor with 16% acceptance. In both cases, we observe that the variance is reduced as the number of games goes up.

***Comparison with Expert Feedback and SizeFlags*** [2]. We ask from at least two fitting experts to reproduce the customer experience and to provide feedback on about 5,000 women shoes (in size 39, which is the average size of the population. The shoes are chosen so that their size 39 is present at least 5 times in the training set). The feedback can be either true to size, too big, or too small. We only focus on SSKUs where all experts agree on the same label. We use SizeFlags [2], a probabilistic Bayesian method to flag articles based on return data as true to size, too big, or too small. SizeFlags do not provide personalized size recommendations like SkillSF, instead they focus on articles. Using SkillSF, we also label an SSKU as too big (resp. too small) when the difference between its label size and its latent size mean after training is larger than 0.2 (resp. smaller than −0.2). Table 5 shows the results for SizeFlags and SkillSF with respect to the Expert Feedback.

SkillSF has higher precision than SizeFlags when the article is too big, though with a lower recall. This is partly explained because SkillSF operates at the SSKU level

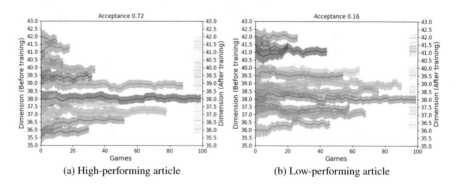

(a) High-performing article                  (b) Low-performing article

**Fig. 3** Changes in articles sizes (Women Shoes)

**Table 5** Expert Feedback versus SkillSF and SizeFlags. The ratios are given with respect to the Expert Feedback

|  |  | SkillSF | | | SizeFlags | | |
|---|---|---|---|---|---|---|---|
|  |  | N/A | Too small | Too big | True to size | Too small | Too-big |
| Expert | True-To-Size | 0.55 | 0.33 | 0.12 | 0.59 | 0.13 | 0.28 |
|  | Too-Small | 0.47 | 0.49 | 0.03 | 0.25 | 0.63 | 0.12 |
|  | Too-Big | 0.48 | 0.05 | 0.48 | 0.30 | 0.10 | 0.59 |

and can isolate each size, while SizeFlags operate at article level and have a higher overview of the article's performance. SizeFlags have better precision and recall for articles that are too small, as also noted in [2], This shows that the experience too big/too small is not symmetric and that a different threshold should be applied on SkillSF to achieve the same precision. It also suggests that return data is overall helpful to determine whether articles have size issues.

## 4.5 Return Data

So far we have trained SkillSF on items the customers have purchased and kept. In this section, we also include during training the purchases returned because the article was too big or too small. Figure 4 shows a drop in acceptance when return data is included of about 5% for Women Shoes and about 2% for Women Upper Garments.

One hypothesis we are investigating is that return data adds instability to the players' skill and can cause radical behavior in the presence of unbalanced distributions

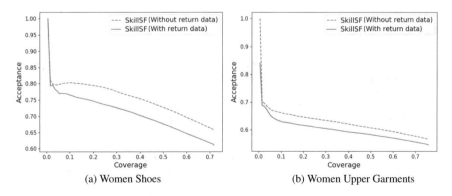

(a) Women Shoes          (b) Women Upper Garments

**Fig. 4** SkillSF trained with and without return data

per SSKU. For instance when an SSKU is mostly returned due to being "too small", the latent size distribution of that SSKU may never converge and end up to far from its actual size.

## 5 Discussion and Future Work

We have presented the problem of personalized size recommendation under the framework of gaming, and introduced SkillSF, a skill ware size recommendation framework that leverages the concept of gaming skills from TrueSkill algorithm [1]. We have successfully applied SkillSF to provide size recommendations. We have shown that this approach is competitive and gives semantically meaningful latent numerical sizes that could be leveraged in downstream tasks such as measurement prediction for both customers and articles, customer segmentation, or article tagging as too big/too small. As future work, we envision two directions to expand the model: First is to train on more categories at the same time to be able to learn correlations between the dimensions of various body parts. Upper Garments and Shoes for example will not have the same set of dimensions, however they communicate via the customers that buy them jointly so SSKUs in Upper Garments can receive information about the average shoe size of their buyers and communicate it to customers who have never bought shoes. Second direction is to add dimensions common to body and garments such as "arm/sleeve length", "waist", "hip", "neck size", etc. as extra skill variables. This will pose the complex challenge of modeling a logic to go from a set of measurements to the prediction of a match between SSKU and customer but could provide invaluable insights.

**Acknowledgements** The authors would like to thank Ralf Herbrich for the support and encouragement given.

## References

1. Herbrich R, Minka T, Graepel T (2007) Trueskill™: a Bayesian skill rating system. In: Schölkopf B, Platt J, Hoffman T (eds) Advances in neural information processing systems, vol 19. MIT Press
2. Nestler A, Karessli N, Hajjar K, Weffer R, Shirvany R (2021) Sizeflags: reducing size and fit related returns in fashion e-commerce. In: to appear in Proceedings of the 27th ACM SIGKDD conference on knowledge discovery and data mining, 2021. arXiv:2106.03532
3. Sembium V, Rastogi R, Saroop A, Merugu S (2017) Recommending product sizes to customers. In: Proceedings of the eleventh ACM conference on recommender systems. ACM, pp 243–250
4. Abdulla GM, Borar S (2017) Size recommendation system for fashion e-commerce. In: KDD Workshop on machine learning meets fashion
5. Sembium V, Rastogi R, Tekumalla L, Saroop A (2018) Bayesian models for product size recommendations. In *Proceedings of the 2018 world wide web conference*, WWW '18, pp. 679–687

6. Misra R, Wan M, McAuley J (2018) Decomposing fit semantics for product size recommendation in metric spaces, p 10
7. Dogani Kallirroi, Tomassetti Matteo, De Cnudde Sofie, Vargas Saúl, Chamberlain Ben (2019) Learning embeddings for product size recommendations. In SIGIR eCom, Paris, France
8. Sheikh AS, Guigourès R, Koriagin E, Ho YK, Shirvany R, Vollgraf R, Bergmann U (2019) A deep learning system for predicting size and fit in fashion e-commerce. In: Proceedings of the 13th ACM conference on recommender systems. ACM
9. Lasserre J, Sheikh AS, Koriagin E, Bergman U, Vollgraf R, Shirvany R (2020) Meta-learning for size and fit recommendation in fashion. In: SIAM international conference on data mining (SDM20)
10. Hajjar K, Lasserre J, Zhao A, Shirvany R (2020) Attention gets you the right size and fit in fashion. In: Submitted to the 14th ACM conference on recommender systems. ACM
11. Januszkiewicz M, Parker CJ, Hayes SG, Gill S (2017) Online virtual fit is not yet fit for purpose: an analysis of fashion e-commerce interfaces, vol 10, pp 210–217
12. Yuan Y, Huh JH (2019) cloth size coding and size recommendation system applicable for personal size automatic extraction and cloth shopping mall: MUE/FutureTech 2018, pp 725–731
13. ABOUT YOU (2020) The customer based size recommendations are accessible on product detail pages. https://corporate.aboutyou.de/en/
14. Thalmann Nadia, Kevelham Bart, Volino Pascal, Kasap Mustafa, Lyard Etienne (2011) 3d web-based virtual try on of physically simulated clothes. Comput-Aided Des Appl 8:01
15. Surville J, Moncoutie T (2013) 3d virtual try-on: the avatar at center stage
16. Peng F, Al-Sayegh M (2014) Personalised size recommendation for online fashion. In: 6th international conference on mass customization and personalization in central Europe, pp. 1–6
17. Bogo F, Kanazawa A, Lassner C, Gehler P, Romero J, Black MJ (2016) Keep it SMPL: automatic estimation of 3d human pose and shape from a single image. CoRR, arXiv:abs/1607.08128
18. Pavlakos G, Zhu L, Zhou X, Daniilidis K (2018) Learning to estimate 3d human pose and shape from a single color image. CoRR, arXiv:abs/1805.04092
19. Karessli N, Guigourès R, Shirvany R (2020) Learning size and fit from fashion images. In: Springer's special issue on fashion recommender systems
20. Guigourès R, Ho YK, Koriagin E, Sheikh AS, Bergmann U, Shirvany R (2018) A hierarchical Bayesian model for size recommendation in fashion. In: Proceedings of the 12th ACM conference on recommender systems. ACM, pp 392–396
21. Lefakis L, Koriagin E, Lasserre J, Shirvany R (2020) Towards user-in-the-loop online fashion size recommendation with low cognitive load. In: Submitted to the 14th ACM conference on recommender systems. ACM
22. Karessli N, Guigourès R, Shirvany R (2019) Sizenet: weakly supervised learning of visual size and fit in fashion images. In: IEEE conference on computer vision and pattern recognition (CVPR) workshop on FFSS-USAD

# Style-Based Interactive Eyewear Recommendations

**Michiel Braat and Jelle Stienstra**

**Abstract** In this demonstration, we introduce STYLE PTTRNS, a style-based tool for eyewear recommendations. The tool first analyses the facial characteristics of a customer and matches those to eyewear characteristics in a style-harmonious manner. Consequently, the customer navigates through the eyewear collection while interacting with high-level expression attributes such as delicateness and strongness. This article presents the customer journey, our pragmatic approach to the recommendation system using styling expert input to bypass a cold-start problem and implementation details.

## 1 Introduction

Digitalisation and the rise of e-commerce in fashion and eyewear retail have not made it easier for customers to find the right product. Whereas customers could rely on a salesperson or stylist advice in a retail context, e-commerce often replicates the product catalogue leaving the customer to browse an abundance of products and discover eyewear that suits based on a picture. Despite Virtual-Try-On solutions, e-commerce faces high rates of product returns as customers are left without style advice and have little support experiencing how the product will actually look on themselves. E-commerce product catalogues benefit from effective filters that help customers scaffold their options as part of their decision-making process but delegates the style, i.e. what is beautiful and works well with, for example, one's facial features and personality, to the customer. We believe that interactive recommender systems—that take a digital role as personal stylists—help build customer confidence, suggest the right product and consequently reduce returns.

Most commercially applied recommender systems use customer demographics such as age, gender and buying history to find similarities between customers in order to suggest products (i.e. others also bought). Others, like iCare [2], use customer feedback and content-based filtering to recommend eyewear. We depart from

M. Braat · J. Stienstra (✉)
PTTRNS.ai, Marconilaan 16, Eindhoven, The Netherlands
e-mail: jelle.stienstra@pttrns.ai

© The Author(s), under exclusive license to Springer Nature Switzerland AG 2022
N. Dokoohaki et al. (eds.), *Recommender Systems in Fashion and Retail*, Lecture Notes
in Electrical Engineering 830, https://doi.org/10.1007/978-3-030-94016-4_5

the perspective that both the eyewear domain, people and context are intricately more complex. For eyewear recommenders to provide highly relevant and personal advice, we consider style as a complex interplay between product aesthetics captured in the product characteristics, customers' facial characteristics, attitudinal preferences like brand sensitivity, trend sensitivity and various dimensions of self-expression. Data-driven recommenders focussed on style cannot easily rely on the sparse prior data and require novel data sets to direct algorithms from an aesthetical perspective. Customers might know what they like, but they are ill-equipped to assess what is beautiful and works for them compared to experts as stylists would. From a commercial perspective, the challenge of a cold start must be overcome to deliver relevant recommendations quickly.

In this industry contributed demonstration, we present STYLE PTTRNS as a personal- and style-based recommender tool for eyewear e-commerce. For an overview of our work, watch the video of the demonstration via https://vimeo.com/585261914. The solution starts with a brief tutorial to ensure customers make a selfie to be processed and analysed by the recommender system. Not unlike eyewear stylists in retail, STYLE PTTRNS uses this initial impression of the customer to suggest the first set of six frames based on expert-defined design rules that match frames that are aesthetically in harmony with the customer's facial characteristics. The recommended eyewear is presented on top of the selfie taken before as a form of Virtual-Try-On. After the initial suggestions, customers can get new recommendations by engaging with sliders that encode high-level expression attributes (e.g. delicateness versus strongness, reserved versus outspokenness, feminine versus masculine, timeless versus trendy), which influence the style of the recommended eyewear. In other words, customers can navigate a narrowed eyewear collection that first matches harmoniously (from an expert perspective) and second considers the customers' expression preference. Rather than letting customers browse through a product catalogue, customers are guided in their choice. The demonstrator focuses on the harmony match and delicateness-strongness dimension of expression.

## 2 Expert-Based Recommendation and High-Level Expression Attributes

STYLE PTTRNS recommender system is based on design rules co-developed by styling experts instead of using historical customer data with methods like collaborative filtering- or content-based filtering techniques. The design rules and recommender came about in a four-step development process. In the first step, initial harmonic recommender design rules were distilled during a 2-hour workshop with several eyewear stylists and a stylist trainer (Fig. 1). This resulted in a series of basic rules for matching customer facial characteristics to eyewear characteristics. For instance, an eyewear frame looks in harmony with someone's face when the shape of the eyewear top corresponds to the eyebrows and that frame colour and contrast to the face should align with the contrast the skin has with features such as hair

**Fig. 1** Demo components walkthrough: A welcome page to STYLE PTTRNS introduces what customers can expect. This is followed by a tutorial for making a good selfie to ensure valuable facial feature extraction. After the customer makes a selfie, face feature extraction feedback shows some of the more visible features like brow shape and colours. Six frames are recommended that are in harmony with the customer's facial features. The eyewear is visualised on top of the customer's selfie. In the last step, customers can interact with the delicateness-strongness expression slider, changing the recommendations accordingly

eyebrows and eyes. In the second step, those rules were implemented in algorithms, after which they were discussed with the stylist in the third step leading to minor adjustments and tuning. In the fourth step, the rules were tested against a validation set labelled by the stylists. Those steps were repeated for the high-level expression attribute delicateness-strongness.

Developing the recommender in conjunction with styling specialists brings about a few advantages. Firstly, there is a close relationship between the physical face features of a person and the aesthetical fit of a frame from a design and styling perspective. This makes methods where a person's facial features are taken into account essential for good frame recommendation. Secondly, another significant advantage is that this alleviates cold-start problems, with no information about a product or customer (who rarely purchases frames regularly compared to other products). By having the customer share information by making a selfie and extracting visual features from it automatically, the customer's effort to input information about itself in our system is minimal. And lastly, using expert rules bypasses the need for (historical sales) data, i.e. customer purchase data, profiles or physical features to train models.

Besides these advantages using expert views for recommendations also gives us an opportunity to look at how personal stylists or style experts at eyewear stores recommend frames and learn about recommendation approaches that are valuable for customers. Customer considerations in a professional stylists recommendation process can be grouped into personal features (body shape, colour, age, body issues) and lifestyle features (profession, daily activities, personality, style preferences) [1]. While a correct or harmonious fit with a customer's personal feature is something the expert can deduce by analysing the customer's physique, the lifestyle considerations are not that easily matched to products. These considerations consist of the person's style preferences, the context in which the customer will be using the product, and all the modalities that come with that. Stylists find and use these lifestyle considerations during the consultation and recommendation process. If a digital counterpart had to gather this information beforehand, this would result in a long, tedious and demanding process for the customer.

Instead, in the STYLE PTTRNS solution, finding the right frames for the customer's specific context and style preference is done after the initial recommendation is given. This is done by empowering the customer to change the recommendations based on high-level expression attributes. For example, if a customer wants eyewear that makes the customer look strong, the customer can use the slider to get recommendations that fit in this style better. This is done by keeping the harmonic fit of the frames in mind. By manipulating high-level expression attributes like the strongness or delicateness of frames, the customer can search relevant parts of the product catalogue on a meaningful level for the customer.

## 3   Implementation Details

The recommendations of STYLE PTTRNS are generated in a multi-stage process. First, we extract customer- and product-specific features. These features are then used in specific rules that encode specific expert considerations based on either the style-harmony or style-expression attributes. The outcomes of these rules are then combined to get the recommendations for the style-harmonious score and the expression style attribute scores. Initially, the recommendations are only affected by the harmonic style score. However, the expression style attribute scores are also weighted into the ranking by changing the attribute sliders. This way, the customer can interactively change the recommendations. The recommendation pipeline is shown in Fig. 2. The following parts further discuss the specific techniques used in the feature extraction and rank calculation based on these rules.

### 3.1   Customer and Glasses Feature Extraction

For the customer feature extraction, we start with the selfie. First, the selfie is validated using the Microsoft Face API. The selfie is checked on the position and rotation of the face, the lighting quality, and whether accessories like glasses do not obscure all parts of the face we want to extract. We make use of a BiSeNet [5] to segment the selfie in different face parts. From this segmentation mask, different shape and colour features are extracted. Colour features are extracted by clustering the pixel colours in the segmentation mask of the specific face part using K-means clustering of the colours in the CIELAB colour space. With these extracted colours, other colour features are extracted, like value and colour contrasts. Shape features are extracted by measuring pixel sizes of the shapes of segmentation masks. These are then normalised based on the distribution of these measurements in an internal selfie data set.

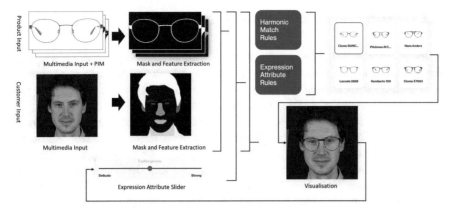

**Fig. 2** The pipeline of STYLE PTTRNS recommendations. The multimedia input for the glasses and the selfie from which features are extracted. Scores for the harmonic style match and the expression attributes are calculated using the features. These are combined in a recommendation and visualised on top of the taken selfie. Using the expression attribute slider, the score of the expression attribute will change and with it the recommendations

**Table 1** These features are extracted from the selfie and the front images of frames that are complemented with product information data commonly used in eyewear e-commerce

| Face features | Eyewear features |
|---|---|
| Face contrast | Frame colours |
| Eyebrow line | Top frame line |
| Chin line | Bottom frame line |
| Eyebrow thickness | Top frame thickness |
| Hair colours | Colour contrast |
| Skin colours | Colour smoothness |
| Lip colours | Frame colour value |
| Undertone skin colour value | Colour temperature |
| Iris colours | Product material |

For the glasses features, we start with a front-facing image of the glasses and product information data. A U-Net [4] was used to segment the frame out of the glasses image, removing the background, legs and glasses. This frame mask is then used to extract colour and shape features in the same manner as is done during the facial feature extraction.

A list of the extracted features can be found in Table 1. Not all possible measures are used in the recommendation algorithm, but only the ones that are important for the style rules developed with the style experts.

## 3.2  Harmonic and Style Attribute Ranking

In contrast to Gu et al. [3] where these rules are implicitly generated, our recommendations are based on explicitly encoded style expert rules developed in cooperation with styling experts using the process described in the previous section. For this demonstrator article, we elaborate on one of these rules.

In the workshop with the style experts, it was distilled that an eyewear frame looks in harmony with someone's face when the shape of the frame top corresponds to the shape of the eyebrows. This rule was implemented in the following algorithm: (1) extracting the lines of the top side of the frame and the underside of the eyebrow of the customer from the multimedia input, (2) normalising these lines, (3) interpolating points on these lines and vectorising them, and (4) calculating the l2 distance between these two vectors. Using this implementation of the rule, distances were calculated between a set of glasses and a set of selfies which were then visualised and validated with the style experts.

Using a set $R$ of these extracted style rules basic function can be created to calculate a score for the harmonic match $H$ between a face $f$ and a pair of glasses $g$:

$$H(f, g) = \sum_{r \in R} w_r * r(f, g) \tag{1}$$

For each customer, we then order the frames by these scores to find the best fitting ones. For each rule, a weight $w_r$ was found based on the expert's feedback on given recommendation rankings for a set of frames on a set of customers. In a similar manner to Eq. (1), scores are calculated for each high-level expression attribute $a$ based on the frames and the customers face features.[1] The function for calculating the scores for other attributes consists of different rules that relate to that specific expression attribute. We combine the harmonic score with the scores for the different expression style attributes and the slider inputs in the following manner:

$$score(f, g) = H(f, g) + \sum_{a \in A} s_a * a(f, g) \tag{2}$$

where $s_a \in [-1, 1]$ is the slider value given as input by the customer for that expression attribute $a$.[2] Each eyewear frame is then ranked based on these scores, and the best scoring ones are displayed based on both the harmonic fitting score and the attribute inputs.

---

[1] In this demonstration, only the strongness-delicateness expression attribute is emphasised.
[2] In the initial recommendation, all sliders are set to 0, removing the impact of the expression attributes.

# 4 Discussion

First customer responses, expert reviews and result comparisons to our expert test set (1k face-frame style-labelled combinations) indicate that our harmony match algorithms and interaction with delicate-strong expression perform well. We need to conduct further quantitative analysis with customers on both the recommended results and the implications of the interaction sliders on customer confidence and, consequently, return reduction.

We seek to improve the harmonic match algorithm further by transitioning from a mainly rule-based method to a data-driven one. We plan to use our new data set of 10k face-frame style-labelled combinations to apply supervised learning methods to the problem of style matching. This way, we will be able to find the stylists explicit rules and the implicit rules they use when assessing whether eyewear matches a customer's face harmonically.

At the moment, STYLE PTTRNS recommends about six frames per customer (and more in groups of six when the customer engages with the expression slider). The eyewear collections of about 3500 frames of mainstream eyewear retailers are scaffolded to a smaller set to prevent recommending six almost identical frames. We seek to resolve high similarity rates algorithmically and offer a more varied recommendation with more significant amounts of frames.

We will continue to expand our style expert approach towards high-level expression attributes like reservedness versus outspoken. And explore how low-level feature interactions like colour, shape and frame thickness impact customers to search through collections from a style perspective. Besides this, future research will also include how we can encode season-related style attributes like timelessness versus trendiness in a sustainable manner and how to incorporate brand sensitivity in the recommendations, as these are both important parts of the attitudinal preferences of the customer.

# 5 Conclusion

In this demonstration of STYLE PTTRNS, different customers can experience a personalised eyewear recommendation that makes sure facial features are matched to those of frames in a harmonious manner. Furthermore, they can navigate through an eyewear collection to find frames that still consider their facial features and take delicateness or strongness expression into account. This demonstration shows how our expert recommendation system attunes to personal preferences in an interactive dialogue. STYLE PTTRNS aims to provide personalised style advice from an expert recommender system.

# References

1. Dahunsi BO, Dunne LE (2021) Understanding professional fashion stylists' outfit recommendation process: a qualitative study. In: Dokoohaki N, Jaradat S, Corona Pampín HJ, Shirvany R (eds) Recommender systems in fashion and retail. Springer International Publishing, Cham, pp 139–160
2. Doody J, Costello E, McGinty L, Smyth B (2006) Combining visualization and feedback for eyewear recommendation. In: Sutcliffe G, Goebel R (eds) Proceedings of the nineteenth international florida artificial intelligence research society conference, Melbourne Beach, Florida, USA, May 11–13, 2006. AAAI Press, pp 135–140
3. Gu X, Shou L, Peng P, Chen K, Wu S, Chen G (2016) Iglasses: a novel recommendation system for best-fit glasses. In: Proceedings of the 39th international ACM SIGIR conference on research and development in information retrieval, association for computing machinery, New York, NY, USA, SIGIR '16, pp 1109–1112. https://doi.org/10.1145/2911451.2911453, https://doi.org/10.1145/2911451.2911453
4. Ronneberger O, Fischer P, Brox T (2015) U-net: convolutional networks for biomedical image segmentation. In: Navab N, Hornegger J, Wells WM, Frangi AF (eds) Medical image computing and computer-assisted intervention - MICCAI 2015. Springer International Publishing, Cham, pp 234–241
5. Yu C, Wang J, Peng C, Gao C, Yu G, Sang N (2018) Bisenet: bilateral segmentation network for real-time semantic segmentation. In: Ferrari V, Hebert M, Sminchisescu C, Weiss Y (eds) Computer vision - ECCV 2018. Springer International Publishing, Cham, pp 334–349

# Fashion Understanding

# A Critical Analysis of Offline Evaluation Decisions Against Online Results: A Real-Time Recommendations Case Study

Pedro Nogueira, Diogo Gonçalves, Vanessa Queiroz Marinho,
Ana Rita Magalhães, and João Sá

**Abstract** Offline evaluation has a widespread use in the development of recommender systems. In order to perform offline evaluation, an Information Retrieval practitioner has to make several decisions, such as, choosing metrics, train–test split strategies, true positives, how to account for biases and others. These decisions have been debated for many years and they are still open to debate today. In this work, we will trial and discuss different decisions that can be taken during offline evaluation for recommender systems. We will then compare their outcome against the results of AB tests performed in an e-commerce production system, which we consider to be the gold standard for evaluation. This is done to verify in an empirical manner which decisions present corroborate with the outcome of the results of an AB test.

## 1 Introduction

The luxury fashion online sales domain has been growing at an accelerated pace throughout the past years. In the online setting, shoppers are able to buy items from any given brand or boutique from any point of the globe in a seamless and swift manner, which reflects the luxurious experience provided in a brick and mortar setting. Farfetch operates in this space and has the mission to bring together creators, curators and consumers of fashion, all over the globe. It offers the biggest catalogue

P. Nogueira (✉) · D. Gonçalves · V. Q. Marinho · A. R. Magalhães · J. Sá
Farfetch, London, UK
e-mail: pedro.nogueira@farfetch.com

D. Gonçalves
e-mail: diogo.goncalves@farfetch.com

V. Q. Marinho
e-mail: vanessa.marinho@farfetch.com

A. R. Magalhães
e-mail: ana.magalhaes@farfetch.com

J. Sá
e-mail: joaomario.sa@farfetch.com

73

N. Dokoohaki et al. (eds.), *Recommender Systems in Fashion and Retail*, Lecture Notes in Electrical Engineering 830, https://doi.org/10.1007/978-3-030-94016-4_6

for luxury fashion in the world, having more than three million products and more than ten thousand brands and high-end designers.

In order to be successful in the luxury fashion space and to cater to the clients' expectation of a luxury-level service, one needs to provide a tailored, personalized and authoritative shopping experience. Therefore, it is vital to have recommender systems that provide the right items to the educated and demanding user of this segment. These systems not only need to be built according to the state-of-the-art and good engineering practices but also require a solid investment on experimentation. This effort is crucial for it adapts the state of the art to a specific audience of a business, guaranteeing that we are able to track down improvements in the user experience as the product evolves.

AB testing is the most common approach to enable Information Retrieval (IR) systems to evolve in a controlled manner [16]. It consists in running randomized field trials on identical sets of users, allowing for the computation of engagement, or satisfaction metrics. IR practitioners are able to make sound and informed decisions of what is the best alternative to present to users by employing hypothesis testing over their feedback. A predetermined Key Performance Indicator (KPI), such as click through rate (CTR) or conversion rate (CR), is commonly used as the metric to define the winner, hence, making a step forward in the IR system. On the one hand, AB tests are done using direct feedback from users, and thus allowing a clear and unambiguous overview of the IR system performance [11]. On the other hand, AB tests have drawbacks to consider [10, 26]. There is a cost associated with collecting real users' feedback, since a considerable share of users might experience a less performant system. Moreover, AB tests need to run for some time in order to determine the winning option with statistical significance. In order to minimize the impact on the final users and not hurt the business' goals, the options AB tested need to be reasonably performant. Due to these reasons, it is paramount to have a method to select the best options to be AB tested. This will allow for a higher rate of successful AB tests, a faster evolution of the platform, and happier customers.

Offline evaluation is typically the best answer for the problem of selecting a short-list of alternatives to be AB tested. Historically, in recommender systems research, the most common methodology consisted of evaluating how well systems could predict the user–item rating from explicit feedback [16]. Yet, nowadays, the focus has shifted towards evaluating users' personal preferences using recommenders built with implicit feedback drawn from users' product interactions within the platform [15, 25]. Commonly, the procedure for offline evaluation of these systems follows the same approach as the classical supervised learning methods, in which one splits the data into train and test. Although the use of offline evaluation is widespread both in academia and industry, several authors claim that the results of these methods are sometimes contradictory when compared to AB test results [9, 16]. These results might also vary depending on the choices taken for offline evaluation [4, 19] as, for example, the decision on how to split the dataset into train and test [19].

The main purpose of this work is to investigate how some of the decisions that are taken in offline evaluation approaches can affect the expected outcome of an experience. We will also observe which of those decisions lead to outcomes that corroborate with the performance indicated by an AB test in a production environment and those that don't. We also present the options AB tested and show the importance of using real-time feedback provided by users in recommender systems.

In summary, our contributions are the following:

1. We empirically experiment how different approaches for computing metrics corroborate with the outcome of AB tests;
2. We test how sampling on the test set might affect the results of offline evaluation;
3. We evaluate if different strategies for choosing true positives corroborate with the outcome of an AB test.

This paper is organized as follows: Sect. 2 refers to related work, Sect. 3 discusses the methodology taken in order to execute the experiments. Section 4 presents and discusses the results obtained in both the offline and online settings. Section 5 outlines the final remarks and, finally, Sect. 6 presents future work.

## 2 Related Works

The study of evaluation of recommender systems has been a crucial and intricate part of the information retrieval field. Early on, Herlocker et al. [14] presented a rigorous overview of many of the aspects that form an evaluation strategy. In their work, topics such as online and offline evaluation, target item sampling and train–test data split are mentioned. Despite being published more than 15 years ago, this work is still relevant today and has paved the way for further research into these matters. Recently, Cañamares et al. [5] also debated the various decisions one can take when designing an offline experiment. In their work, they discussed several issues from the options available for the train–test split decision, to the items used for prediction in the test set, revealing that some of these choices, such as the train–test split, might affect the final outcome of an offline experiment. Jeunen [16] has also performed a good overview on some of the relevant decisions made for offline evaluation in recommender systems. The author addressed topics such as temporal evaluation and debiasing logged user interactions. Ludewig et al. [20] performed an extensive review of offline evaluation in the particular context of session based recommenders where they benchmarked several algorithms using a defined set of decisions for evaluation. Felicioni et al. [8] proposed a novel approach for the specific scenario of offline evaluation for a carousel user interface which resembles a production interface with several carousels. The authors said that the quality of a model should not be measured independently but it should take other recommendation lists into account.

The industry is also applying the best practices regarding offline evaluation and in some scenarios it appears to corroborate with online performance. This can be observed in the work of Pal and Eksombatchai et al. [23], where the authors presented a method used by Pinterest for retrieving information for the user. The results revealed that the offline experiments performance pointed towards the same direction as the online ones. The corroboration between offline and online evaluation was also verified in the work presented by Yi et al. [27]. Gruson et al. [12] also showed a corroboration between offline and online performance of algorithms by using several AB tests as a golden standard for evaluation and then compared the results against the offline evaluation outcome. However, just as Jeunen [16] and other authors [2, 9] mentioned in their work, this corroboration does not always occur.

Cañamares et al. [4] showed that different decisions for the target size configuration for offline evaluation might lead to different results and, thus, the community should be cautious and rigorous in taking such decisions. Krichene et al. [19] argued that the long-used practice of using sampled metrics [7, 13] was wrong in most scenarios, and that this enduring procedure should only be used under specific circumstances.

The data used by offline evaluation methods also suffers from several biases that might affect the outcome of their results. Chaney et al. [6] discussed this matter in their work, where they stated that algorithmic confounding increases the homogenization of user behaviour and changes the distribution of item consumption. Thus, researchers should be aware when using confounded data for user behaviour studies. Joachims et al. [12] first studied this matter by analysing position and presentation bias for training models and found that the position in which the clicked item was placed significantly altered the capability of these models to estimate the relevance of a document. Bottou et al. [3] discussed methodologies, such as inverse propensity scoring (IPS), to extract signals from biased data and how to create algorithms to learn from it.

In summary, we debate the topic of the decisions that can be taken during offline evaluation which is heavily mentioned in the literature. We hope that by experimenting with some of those different decisions, such as the usage of sampled metrics or different methods for the selection of true positives we might be able to add relevant research to this debate.

## 3   Methodology

### 3.1   How to Do Offline Evaluation

Offline evaluation of recommender systems is a topic that has been studied and debated for some time [14]. On this very day, despite having some clarity and consensus on this topic, there are still some open issues [4, 16, 19]. These issues are, for example, the items to be used during the prediction step of the evaluation, or the ratio used on train–test split. There are different decisions during the offline evaluation process that could eventually lead to different results [4, 5, 19].

In this section, we discuss a shortlist of decisions that need to be taken during offline evaluation and which are still open to debate in both Academia and Industry.

### 3.1.1  Creating the Test Dataset

The train–test split methodology can be done in several manners. The common approach of random splitting of data in train and test, which serves well the supervised learning applications, does not apply to recommender systems because there is a strong temporal dependency of data [17]. However, the method for conducting the split is not universally accepted across Academia and Industry, leaving room for discussion. Therefore, we pose the question of how one should split the dataset into train and test in recommender systems.

Cañamares et al. [5] suggested different strategies for these methods and observed those could lead to opposite results. According to these findings, an IR practitioner should be aware of the influence the train–test split might play on the results, in order to make considerate decisions during this process.

One of the options used for evaluation that has taken traction [16, 20] is the temporal leave-one-out methodology introduced by Zhao et al. [28]. Jeunen et al. [17] builds upon this methodology and proposes the "Sliding-Window Evaluation" method that provides a robust estimate by following the chronological order of the interactions. In this method, the test set will be composed by the last item the user interacted with in a given moment. This ensures that no future preferences influence the recommender at the time of the recommendation, therefore mimicking a real life scenario.

We opted to follow the Sliding-Window Evaluation methodology [17] and to use the temporal leave-one-out method while iterating over the dataset in a chronological order. After a specific event has been used for testing, we use it to update our recommenders so that they can accommodate new information for the following event. This logic is illustrated in Fig. 1. By doing this, we attempt to mimic what happens in our system in production [23].

E-commerce platforms generate massive amounts of information, having millions of users interacting with the websites every day. The volume of this data can impact the speed in which IR practitioners experiment. In order to diminish the experimentation time and speed up the innovation process, one can sample the users that compose the test set. Yet, by conducting sampling, the original distribution of data might change, possibly leading to different outcomes in offline evaluation. Therefore, we will experiment with two different settings regarding this decision. In one

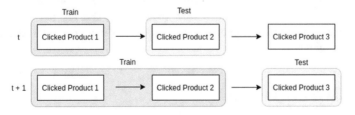

**Fig. 1**  Example of sliding-window evaluation

setting, all observed users will be used to build the test set, and, in the other one, we will randomly sample 50% of the users in the test set. We will then observe the results to determine if both decisions lead to the same result.

### 3.1.2 Selecting the Candidates for Prediction

Any recommender system requires an input of user–item pairs in order to output a score to obtain personalized recommendations. Given this, we pose the question of what set of items should the recommender system order when computing offline evaluation metrics?

In the past few years, it has been fairly common to find the use of sampled metrics in the literature [7, 18, 20] in order to accelerate the computation of metrics. Sampled metrics is a term used to reference the metrics calculated in a subset of all the possible candidates for recommendation, which should include relevant items and a sample of randomly selected ones. Then, recommender systems are given the task of ordering the selected sample. The usage of sampled metrics has been open to debate [4, 19] with researchers posing contradictory arguments.

Krichene et al. [19] stated that sampled metrics approaches may lead to misleading results. They argued that sampled metrics do not provide comparable estimates with exact versions of the metrics. In their work, the authors mentioned that these metrics can be viewed as high-bias and low-variance estimators of the exact metrics. Therefore, they warned practitioners that the usage of such an approach should be avoided and proposed a correction that can improve the quality of the estimates produced by these metrics.

Cañamares et al. [4] stated that the amount of random items added during the calculation of sampled metrics might impact the final outcome of the experiment. In order to prove this, the authors gradually varied the target size in several experiments and attempted to find a sweet spot between using a small sample of the data or using all of it. The authors also mentioned that using the full data may lead to algorithms converging to a global tie, causing metrics to converge towards zero. This occurs because finding a relevant item in the midst of millions of other items is a very challenging task for any recommender system. Finally, they stated that avoiding the tie phenomenon is the main advantage of using sampled metrics over the full data set, since there are less items in need of ordering.

This debate makes this issue relevant for experimentation and, thus, we opted to experiment both methods and observe their impact on the final results. By making this decision, we hope to obtain results that will point us towards an optimal configuration for offline evaluation.

### 3.1.3 Selecting True Positives

Defining the true positives to be used during experimentation is also a relevant decision. First and foremost, one can decide to use implicit or explicit relevance judgements based on the collected users' feedback [11]. Although being the most obvious and, probably, the most reliable source of relevance, explicit feedback is not easy to collect on most platforms and applications. User–item ratings built using explicit feedback are sparser for they depend on users to rate a fair number of products. This might be the reason why such an approach has been somewhat disregarded in favour of the implicit feedback.

In an implicit feedback setting, it is not obvious what relevance means and how it translates to user satisfaction and engagement. The most common proxies for judging the relevance of an item to a user in e-commerce are clicks, add to wish lists and purchases. In this work, we will focus on working with implicit user–item relevance judgements.

Implicit feedback approaches present some problems to practitioners, raising questions regarding the assignment of weights to each implicit action. For example, answering the question of 'how much is a click "worth" when compared to a purchase?' is not obvious. For the sake of simplicity, we chose to use clicks as relevance feedback, since it is not straightforward to define the weights for each action in a robust manner [27]. This will also mirror the metric (CTR) which was used to determine the winner of the conducted AB tests, thus making the offline evaluation theoretically comparable to the online one.

Selecting which clicks should be defined as true positives is also a decision worth discussing. For example, should we only consider the clicks from the specific carousel (Fig. 2) where the recommendations are presented? Or should the clicks of the entire user session be considered as positive feedback? We believe that both options make sense in the light of offline evaluation. The first approach focuses on the specific carousel where the different AB test alternatives will be displayed and thus, the final results of the experiment should theoretically be closer to the ones from the AB test. Yet, the second approach better reflects the overall users' intentions, for it considers all clicks in the whole platform in a given session. Therefore, the offline evaluation results might be able to reward the alternative that better suits the users' needs. With this in mind, we pose the question of how to select true positives for offline evaluation?

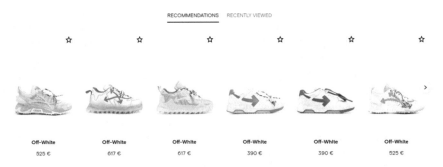

**Fig. 2** Example of recommendations carousel

To answer this, we decided to evaluate and report results using three different options for the selection of true positives, as listed below:

- Option 1—True positives are the clicks done by a specific user in a session;
- Option 2—True positives are all the clicks done in a session grouped by the specific context of the event. Context can be any combination of gender, brand and category;
- Option 3—True positives are the specific clicks in the recommendations carousel.

### 3.1.4 Metrics

We opt to observe several IR metrics, for each one might provide additional information that another one does not. This will allow us to have a fuller picture of the final outcome of the experiment. In addition, we might find corroborations between the outcome of the online trial and a given metric. We use the following metrics with a cutoff of 15.

- Mean Reciprocal Rank (MRR) @15;
- Precision@15;
- Recall@15;
- Hit Rate@15;
- Normalized Discounted Cumulative Gain (nDCG) @15.

### 3.1.5 Biased Feedback

The data used in this work was obtained by logging the user–item interactions in an e-commerce platform. These were generated by logging the rewards (clicks) of a user, based on the recommendations provided by a recommender given a user and a specific context (brand, category or gender). We use the terminology of a multi-armed bandit framework and state that a recommender follows a policy distribution by drawing items stochastically conditioned on context [21].

The logs produced with a determined policy suffer from bias because they were obtained by the users interacting with an existent information retrieval system [16]. These biases can then cause feedback loops because a new recommender might be rewarded by predicting the output of the existing recommender instead of focusing on the task of providing relevant items to the user. This problem hinders the offline evaluation methodologies and possibly might even lead the IR practitioners to make incorrect decisions.

One of the solutions to counter the bias produced by existing recommenders is inverse propensity scoring (IPS) [22]. This method attempts to estimate the average reward of a new policy using data collected from the existing policy while correcting the inconsistencies between both policies. Due to a vast amount of items present in the catalogue of e-commerce platforms, IPS estimators might present some near zero propensity weights and thus suffer from high variance [10]. A solution to this problem is propensity clipping. In this method, one clips the propensity scores as they reach a certain value in order to decrease the variance of the estimation. However, this method leads to an increase of bias [10].

In most e-commerce platforms, there are several ways for users to observe items. This is not only because of the maturity of these platforms but also because of several different systems that power them. For example, a user might interact with various recommender systems during a single session. These various sources decrease the bias because a user is not only interacting with one single system but with many. This is also one of the reasons for the decisions made in this work regarding the selection of true positives mentioned in Sect. 3.1.3. For example, by considering all clicks a user made in a session (not only those from recommendations) as true positives, we are severely decreasing the effect that the existing policy for a single source of data might have in the final outcome of offline evaluation. One of the options considered for true positives (option 3 of Sect. 3.1.3) will suffer from more biases than two other referred options. Yet, it is relevant to mention that more than one policy is used to retrieve items for that carousel. Due to having a diverse set of policies for retrieving items and because of having near zero propensity values, we opted not to use IPS in this work.

### 3.1.6  Design of Experiments

Several experiments will be conducted in order to trial different offline evaluation decisions. Below, we present a summary of all of the decisions that will be trialled:
    Selecting True Positives:

- True positives are the clicks done by a specific user in a session;
- True positives are all the clicks done in a session grouped by the specific context of the event. Context can be any combination of gender, brand and category;
- True positives are the specific clicks in the recommendations carousel.

    Selecting the candidates for prediction:

- Relevant items for the user shuffled with one hundred random items;
- All items available at the time.

    Creating the dataset:

- Using all the users in the test set;
- Randomly selecting 50% of the users for the test set.

These different decisions will be assessed in a combinatorial manner, effectively translating into 12 different experiments per recommender.

## 3.2  Dataset

In order to have a better understanding of the experiments performed, it is relevant to explain the data used in those. Two different datasets were used in this work, a train and a test dataset. This might seem somewhat atypical for a machine learning practitioner, yet the reasoning behind this decision and why, in this scenario, this leads to a more rigorous evaluation protocol will be presented below.

The first dataset is the one used to train the recommenders and it is composed of several matrices. The first step necessary to obtain these matrices is to build a user–item ratings matrix. We build this matrix by attributing a weight to each user–item action (click, add to bag, purchase, etc....). A decay is then applied to each one of these actions according to how much time has passed since it was performed. After the decay is applied, we combine all the user–item actions and obtain a matrix $U_r$. This matrix contains user–item ratings that should represent users' affinity towards items. Another necessary matrix for the train dataset is the matrix $I_f$ that contains all the products and their features, such as brand, attributes, category or others. By multiplying $U_r$ by $I_f$, we obtain $U_f$. This matrix holds a representation of users according to the features of the items they interacted with. The three matrices $U_r$, $I_f$ and $U_f$ effectively make up the train dataset required to train the recommenders presented in this work. It is also relevant to note that all of them are computed in a daily fashion to keep up with the new user interactions. These matrices mirror the ones used by the recommenders in the production system. Therefore, by using them as train data, we are able to obtain an exact image of the recommender in production on a certain day.

The second dataset is what we consider to be the test dataset. This dataset is composed of the clicks the users performed on the platform and their contextual information, such as the type of page in which it occurred or the clicks performed previously. With this information, we are able to recreate the exact moment in which a user clicked on an item, therefore having a more realistic offline evaluation. The test dataset was built using several days worth of data and contains information about 450 thousand unique products, 750 thousand clicks and 120 thousand users.

By having two datasets, we are able to (1) recreate a recommender for a determined day as it was in the production system and (2) perform offline evaluation of the users' clicks in a more exact manner, by considering each individual event a user session contains.

The purpose of our offline evaluation is to determine how the models trained with the train dataset performed in ranking the items for each click event in the test dataset. We purposely chose not to use public datasets because the goal of this work is to compare the outcome of offline evaluation against the results of an AB test. To the best of our knowledge, there aren't any public datasets containing this information.

## 3.3 Experimentation

In this work, three different recommender systems were tested in an online and offline fashion. Two of these recommenders work in a very similar manner, following a contextual hybrid approach [1], and the third applies the well-known and effective methodology of Factorization Machines [24]. We selected these widely used recommenders because the scope of our work was not to develop new algorithms.

Context-aware recommenders systems (CARs) [1] try to increase their predictive power by using any sort of context information. Context can refer to multiple aspects: time and location from where the user is navigating, intent/purpose of navigation, the place in the website, among others. In this recommender, we consider context as the combination of a gender of a page (women, men or kids) and the respective brand and/or category of the page where the carousel is shown. In this specific scenario, consider context $c$, a user $u$ and the features that represent an item, such as brand, attributes, category or others, $i_f$. The context-aware recommender selects the products that $u$ has interacted with in order to obtain $u_f$. This is a representation of $u$ in the space of $i_f$. Similarly, all users are represented in the space of $i_f$ in the matrix $U_f$. This matrix holds a representation of users according to the features of the items they interacted with. $U_f$ is then sliced so that it contains all users that interacted with context $c$, thus obtaining $U_{fc}$. The recommender then proceeds to find the most similar users to $u$ by comparing $u_f$ with all users in $U_{fc}$. After these users are found, the algorithm follows a collaborative approach in order to provide recommendations to $u$. It does so by recommending products that the most similar users to $u$ also interacted with. Two of the algorithms tested in this work use this approach. The main difference between them is the frequency in which $u_f$ gets updated. The user representation is updated daily in the batch-like version, while it gets updated in real time in the other approach.

In addition to the algorithm explained above, we also used the recommender Factorization Machines (FM) [24], which is well known by the community and works notably in the recommendation problem. In our implementation of factorization machines, we use $U_f$ mentioned previously as input features. This recommender will attempt to predict a user–item affinity score by grouping the user interactions with items into a single value. This algorithm will also be updated in a real-time manner, therefore the actions taken by users will immediately affect their next recommendation.

We will refer to these methods as such:

- cars hybrid batch—Contextual hybrid approach following a batch-like update;
- cars hybrid real-time—Contextual hybrid approach following real time-like update;
- fm—Factorization machines approach following real time-like update.

# 4 Results

In this section, we will present and discuss the results obtained via AB test and offline evaluation for two different experiments. Both experiments will test the algorithms described in Sect. 3.3.

To guide the discussion in this section, we will follow the 12 combinations of decisions outlined in Sect. 3.1.6. For the online evaluation purposes, we will observe the increment in CTR reported in both AB tests in order to determine how much the offline evaluation agrees with the results obtained via AB test.

For the sake of simplicity, we will refer to the different trials as:

- **Trial A**—The experiment that compares the algorithm cars hybrid against its real-time version;
- **Trial B**—The experiment that compares the real-time version of cars hybrid against the implementation of factorization machines.

The results presented are divided by gender, so that they mirror the production system, which also works based on the gender of the page the users are browsing. This occurs because the platform used to obtain the data has gender inherently built into it. We only present the results for the gender "men" because the ones for the gender "women" always point to the same relative direction as gender "men".

## 4.1 Live Experiment

In the subsections below, we describe the two live experiments that ran on our platform. The product recommendations were placed in a carousel in product listing pages, which are pages that aggregate products according to a criteria, such as same gender, brand and/or category, etc. An example of such carousel can be seen in Fig. 2.

The system in production when trial A was performed was the recommender cars hybrid batch and trial B was executed it was the recommender cars hybrid real time. Yet, it is important to state that a single carousel might present different recommendation strategies besides the mentioned recommenders, depending on the user journey, context of the page, among others.

### 4.1.1 Trial A—Cars Hybrid Real-Time X Cars Hybrid Batch

In this section, we present an AB test that compared two strategies, the control strategy used the algorithm cars hybrid batch, while the alternative used cars hybrid real time mentioned in Sect. 3.3. The traffic was split equally among those strategies. The alternative strategy was the winner, with a positive CTR uplift.

We calculated the CTR uplift using the binomial proportion with a significance of 95% by comparing the alternative with the control for users by gender. To obtain

**Table 1** CTR uplifts for users split by gender for trials A and B

| Trial | Gender | CTR Uplift (LB) (%) | CTR Uplift (UB) (%) |
|-------|--------|---------------------|---------------------|
| Trial A | Men | 0.27 | 9.8 |
| Trial A | Women | 4.7 | 11.8 |
| Trial B | Men | 68.48 | 101.85 |
| Trial B | Women | 43.53 | 63.51 |

the lower and upper CTR uplifts, we first calculated the confidence intervals of the average CTR for the control and alternative and then computed the interval limits. An example of the lower CTR uplift is presented in Eq. 1:

$$Uplift_{lower} = \frac{CTR_{lower}^{Alternative} - CTR_{upper}^{Control}}{CTR_{upper}^{Control}} \tag{1}$$

The CTR uplifts for users split by gender are presented in Table 1. We observe that the CTR uplifts are positive for both genders, and the impact of adding real-time signals is higher for women than men.

#### 4.1.2　Trial B—Cars Hybrid Real-Time X Fm

In this section, we present the second live experiment. In this AB test, the control strategy used the algorithm cars hybrid real time, while the alternative used the implementation of factorization machines. The main success metric assessed during this test was CTR. The alternative strategy was the winner, with a positive CTR uplift.

The CTR uplifts for users split by gender is presented in Table 1. These were calculated in the same way as in Sect. 4.1.1. The CTR uplift of the alternative over control is considerably high. When we compare both genders, the uplift is higher for men than women. One possible explanation is that the set of available products for women recommendations is larger, which makes the problem usually harder for this gender.

### 4.2　Offline Results

In this section, we present the results of the offline experiments conducted in this work in Tables 2 and 3. In order to obtain the values for these tables, we first computed the average of the daily results for each metric mentioned in the Sect. 3.1.4 for both recommenders in both trials. Then, we computed the uplifts comparing one recommender against the other. This process mirrors the methodology used during the AB tests that evaluated the same recommenders. The comparison of the different

**Table 2** Trial A—results for gender men for cars hybrid versus its real-time version

| True positives | Candidate selection | User sample (%) | Hit @15 (%) | MRR@15 (%) | Prec. @15 (%) | Recall @15 (%) | nDCG @15 (%) |
|---|---|---|---|---|---|---|---|
| Carousel specific clicks | Full metrics | 50 | 33.20 | 46.83 | 33.58 | 33.33 | 41.81 |
| | | 100 | 26.08 | 32.34 | 25.92 | 26.53 | 31.44 |
| | Sampled metrics | 50 | 26.57 | 46.69 | 26.35 | 26.65 | 37.10 |
| | | 100 | 22.42 | 37.95 | 21.51 | 22.51 | 32.92 |
| Clicks in session per context | Full metrics | 50 | 58.36 | 64.08 | 61.88 | 63.05 | 62.79 |
| | | 100 | 59.60 | 67.48 | 62.25 | 64.34 | 65.05 |
| | Sampled Metrics | 50 | 17.61 | 25.20 | 21.75 | 21.20 | 22.28 |
| | | 100 | 18.12 | 25.64 | 21.38 | 21.54 | 22.76 |
| All clicks in session | Full metrics | 50 | 43.84 | 53.32 | 50.38 | 58.92 | 49.64 |
| | | 100 | 47.11 | 57.15 | 52.06 | 57.79 | 52.03 |
| | Sampled Metrics | 50 | 2.81 | 8.75 | 9.09 | 13.02 | 5.04 |
| | | 100 | 3.29 | 10.43 | 11.58 | 12.75 | 5.86 |

recommenders is done for the combination of configurations presented in Sect. 3.1.6. All results presented are statistically significant. We obtain significance by using a paired samples t-test, computed with a significance of at least 95%, by comparing the mean of a given metric for both recommenders. We employed a paired samples t-test because both recommenders use the same data in the offline evaluation process.

During this section, we will debate the results and examine how the decisions made affected the outcome of offline evaluation. This will be done by comparing the results of offline evaluation against the ones from the AB tests which we consider to be the ground truth.

### 4.2.1 Creating the Test Dataset

In both trials conducted in this work, we experimented if using a random sample of 50% of the users present in the test set would lead to the same results as using all the users for the test set.

**Table 3** Trial B—results for men for cars hybrid real-time versus factorization machines

| True positives | Candidate selection | User sample (%) | Hit @15 (%) | MRR@15 (%) | Prec. @15 (%) | Recall @15 (%) | nDCG @15 (%) |
|---|---|---|---|---|---|---|---|
| Carousel specific clicks | Full metrics | 50 | −40.00 | −28.70 | −37.85 | −40.67 | −37.67 |
| | | 100 | −43.58 | −36.51 | −42.97 | −44.54 | −40.86 |
| | Sampled metrics | 50 | 25.06 | 58.60 | 19.25 | 25.97 | 43.71 |
| | | 100 | 28.13 | 60.27 | 23.25 | 28.71 | 44.55 |
| Clicks in session per context | Full metrics | 50 | −36.45 | −45.57 | −36.01 | −39.29 | −42.92 |
| | | 100 | −37.26 | −49.36 | −36.91 | −40.33 | −45.21 |
| | Sampled metrics | 50 | 8.14 | 8.21 | 8.60 | 11.28 | 9.30 |
| | | 100 | 7.10 | 5.40 | 7.02 | 10.65 | 7.21 |
| All clicks in session | Full metrics | 50 | −13.54 | −25.04 | −12.36 | −26.94 | −19.92 |
| | | 100 | −9.79 | −20.05 | −8.85 | −24.87 | −16.87 |
| | Sampled metrics | 50 | 4.07 | 13.87 | 16.43 | 13.61 | 8.69 |
| | | 100 | −21.58 | 4.70 | 21.73 | −45.38 | −9.75 |

The results of these experiments can be seen in Tables 4 and 5. In them, we can observe the uplift of a given metric of a certain configuration that uses a random sample of 50% of the users in the test against another version that uses all the users in the test set.

In Table 4, we can observe the results for trial A. We note that only the values for some specific configurations are statistically significant. In the remaining configurations, we obtain no statistical significance. This can occur for three reasons: (1) our sample is not big enough; (2) the distribution of the samples observed are very similar to one another or; (3) both reasons combined. It is interesting to see that in trial A the uplifts reported in Table 4 for all metrics are rather small. We can see that the maximum uplift occurs for Precision@15 with a value of -4.96%. We also observe that there are only significant uplifts when using sampled metrics. Therefore, the values of this trial seem to validate the hypothesis that there is no substantial difference between using a sample of 50% of the users in the test set and all users.

It is interesting to observe that the results for trial B, present in Table 5, yield different results than the ones presented in Table 4. In the results presented in Table 5, we see that several of the metrics are significantly different for some options. There is even a Recall@15 uplift as low as −54%, which is a big difference to the metric

**Table 4** Trial A—results comparing the uplifts in metrics between the same experiment using 50% of users versus all the users in the test set[1]

| True positives | Candidate selection | User sample (%) | Hit @15 (%) | MRR@15 (%) | Prec. @15 (%) | Recall @15 (%) | nDCG @15 (%) |
|---|---|---|---|---|---|---|---|
| Carousel specific clicks | Full metrics | cars hybrid batch | 5.54 | 4.68 | 6.77 | 6.28 | 3.22 |
| | | Cars hybrid real-time | −0.10 | −5.65 | 0.64 | 0.87 | −4.33 |
| | Sampled metrics | Cars hybrid batch | −0.94 | 1.26 | −1.18 | −0.79 | −0.63 |
| | | Cars hybrid real-time | −4.19 | −4.78 | −4.96* | −4.03 | −3.66 |
| Clicks in session per context | Full metrics | Cars hybrid batch | −0.42 | −0.71 | −0.71 | −0.04 | −0.62 |
| | | Cars hybrid real-time | 0.36% | 1.35 | −0.48 | 0.75 | 0.77 |
| | Sampled metrics | Cars hybrid batch | −0.62 | −0.26 | 0.66 | −0.61 | −0.42 |
| | | Cars hybrid real-time | −0.19 | 0.09 | 0.35 | −0.33 | −0.03 |
| All clicks in session | Full metrics | Cars hybrid batch | −1.33 | −2.67 | −0.66 | −0.49 | −1.17 |
| | | Cars hybrid real-time | 0.91 | −0.24 | 0.45 | −1.19 | 0.41 |
| | Sampled metrics | Cars hybrid batch | −0.61 * | −1.48 * | −1.26 * | −2.41 * | −0.88 * |
| | | Cars hybrid real-time | −0.14 | 0.04 | 0.99 * | −2.65 * | −0.11 |

Values marked with a * represent statistically significant uplifts. This significance was obtained using a paired samples t-test computed with a significance of at least 95%

**Table 5** Trial B—Results comparing the uplifts in metrics between the same experiment using 50% of users versus all the users in the test set [1]

| True positives | Candidate selection | User sample (%) | Hit @15 (%) | MRR@15 (%) | Prec. @15 (%) | Recall @15 (%) | nDCG @15 (%) |
|---|---|---|---|---|---|---|---|
| Carousel specific clicks | Full metrics | Cars hybrid real-time | −1.41 | −3.23 | 0.30 | −0.03 | −4.32 |
| | | Fm | −7.29 | −13.83 | −7.95 | −6.55 | −9.22 |
| | Sampled metrics | Cars hybrid real-time | **−6.23 \*** | −6.05 | −6.86 | −6.05 | −4.64 |
| | | Fm | −3.93 \* | −5.07 | **−3.74 \*** | **−4.01 \*** | −4.08 |
| Clicks in session per context | Full metrics | Cars hybrid real-time | −2.76 | 2.50 | *−3.22 \** | *−3.23 \** | 0.05 |
| | | fm | −4.00 | −4.63 | −4.57 | −4.88 | −3.97 |
| | Sampled metrics | Cars hybrid real-time | **−0.99\*** | **−0.89\*** | −2.24 \* | **−1.50\*** | **−0.90\*** |
| | | Fm | −1.94\* | −3.46 | **−3.67\*** | **−2.06\*** | **−2.80 \*** |
| All clicks in session | Full metrics | Cars hybrid real-time | **0.16 \*** | **−1.66 \*** | −0.67 | −2.72 | **−0.25 \*** |
| | | Fm | 4.52 | 4.90 | 3.31 | 0.03 | 3.55 |
| | Sampled metrics | Cars hybrid real-time | **−0.47 \*** | 0.22 | **1.38 \*** | **−4.98 \*** | −0.27 |
| | | Fm | −25.00 | −7.85 | **5.99 \*** | **−54.32 \*** | −17.19 |

Values marked with a * represent statistically significant uplifts. This significance was obtained using a paired samples t-test computed with a significance of at least 95%

observed for all users. Ideally, this should not happen because we are randomly selecting 50% of the users in a very large dataset, yet it does occur. It is also relevant to state that we only observe large uplifts when using sampled metrics. When using full metrics, the biggest uplift is −3.23%. This might happen because the combined usage of sampled metrics and a sampled test set will increase the variance of the final outcome.

In both trials, when using full metrics we do not verify any considerable differences between using a sampling strategy in the test and not using one at all. Yet, when using sampled metrics we note small differences for trial A and substantial differences in trial B. Therefore, we believe that the hypothesis stating that there is no substantial difference between using a sample of 50% of the users in the test set and all users is in fact correct for the usage of full metrics. Yet the combined use of sampled metrics and a sampled test set seem to lead to substantial differences when comparing the usage of a sample of 50% of the users in the test set and all users. This is particularly true in trial B. This mismatch could lead to potentially wrong conclusions, yet, this only happened when we used sampled metrics. Therefore, we believe that carefully defining a sampling strategy for the creation of a test set is particularly important. Due to the differences found in trial B, we believe one should use all users in the test set when using sampled metrics, even though that option might take considerably more time. It also seems that when using full metrics one can use the option of sampling 50% of the users in the test.

### 4.2.2 Selecting the Candidates for Prediction

During our trials, we experimented with different sets of candidate products for the recommendation. In the first approach, we used all available products as the candidates for the recommendation. The second approach consisted of defining a specific set for each user. In this set, we included all the products clicked by the user on the test day plus 100 random products. The main reason behind this decision was not only the time taken in order to rank the candidates but also to compare the full with the sampled metrics.

The results of these experiments can be seen in Tables 2 and 3. In them, we can observe the uplift of a given metric for the configurations experimented for trials A and B. As we mentioned in Sect. 4.1.1, we observed a CTR uplift between 0.27% and 9.8% for the gender men in trial A. This corroborates the uplifts we observed when we used sampled metrics in Trial A. For example, the uplifts in Hit@15 for the sampled metrics are around 20% or lower, while the full metrics have a much higher positive uplift.

The uplifts observed for the sampled metrics in trial B are also more similar to the ones observed in the live experiment, as they are usually positive. Interestingly, full metrics present a negative uplift in this trial. As we had mentioned in Sect. 3.1.2, the literature had suggested that using full data might lead to global ties [4]. This seems to be playing a role in this trial. It's much harder for the algorithms to distinguish a relevant item in the whole list of available products, from a scenario where there are some relevant items mixed with a few random ones. This seems to affect factorization machines more than cars hybrid real-time recommender.

From the results described above, it seems that sampled metrics corroborate in a better way with the outcome of our AB test results. These results corroborate with the work of Cañamares et al. [4], which recommended using sampled metrics as a way to avoid the tie phenomenon. In a different direction, Krichene et al. [19] stated that sampled metrics might lead to misleading results and were not comparable estimates of the full versions of the metrics. It is possible that with a bigger sample size of AB tests, our results would point to the same direction as the work from Krichene et al. [19], yet that question remains open.

### 4.2.3  Selecting True Positives

For both trials A and B, we noted that the different options for the true positives point to the same "direction". For trial A, all the different choices corroborate with the outcome of the corresponding AB test as can be seen in Table 2 and the AB test results presented in Section 4.1.1. For trial B, we can observe in Table 3, that the same happened for all the true positive selection choices for the sampled metrics approach. This is interesting and it seems to validate our hypothesis stated in Sect. 3.1.3. Therefore, it seems that both ideas behind the selection of true positives mentioned in Sect. 3.1.3 make sense, namely (1) choosing only the clicks for a specific touchpoint where an AB test would be conducted (Option 3 of Sect. 3.1.3) and (2) broadening those true positives with other clicks in the session (Option 1 and 2 of Sect. 3.1.3).

No choice for the selection of true positives appears to have provided a good estimate on the exact value of the uplift calculated in each AB test. All the choices apparently overestimate the uplift for trial A, while in trial B the true positive selection choices underestimate the uplift.

## 5  Conclusion

From this work, we can conclude that, in fact, the decisions made during the design of a given offline experiment are paramount, for different decisions might lead to different results. Our empirical findings seem to indicate that sampled metrics corroborate in a better way with the outcome of an AB test, when compared to the full version of those metrics. It is a curious finding considering that Krichene et al. [19] work stated the exact opposite. We also note that the sampling strategies for building the test dataset might lead to different results when using sampled metrics. Therefore, it is important to choose them carefully if IR practitioners intend to use one. We finally also observe that different strategies for selecting true positives yield results that point to the same direction. This indicates that using an approach that does not select true positives from a specific place might be a good strategy to mitigate bias, while still corroborating with the outcome of an AB test.

## 6  Future Work

We believe that there are some interesting topics that could be further developed from this work. Performing these type of experiments is computationally heavy and also requires some time. Due to this, some trials that could provide interesting insights were not carried out. For example, these could consist of experiments with different sample sizes for the random selection of users in the test set, or even different methods of sampling entirely. Experimenting with different proxies to a relevance metric besides clicks is also something that we are looking forward to doing in the future. It would also be very interesting to perform similar experiments to those executed by Cañamares et al. [4] where one would trial several sampling sizes in the sampled metrics process. In addition, it would be valuable to analyse the effect of the different sample sizes for each metric. Executing these experiments using more AB tests is also something that we are looking forward to doing in the future. By doing this, we would increase the sample of AB tests thus having more robust results. In summary, we believe that providing results for empirical experiments and comparing them to the outcomes of AB tests is relevant and helps the community to make better decisions for offline evaluation.

# References

1. Adomavicius G, Mobasher B, Ricci F, Tuzhilin A (2011) Context-aware recommender systems. AI Mag 32:67–80
2. Beel J, Genzmehr M, Langer S, Nürnberger A, Gipp B (2013) A comparative analysis of offline and online evaluations and discussion of research paper recommender system evaluation. In: Proceedings of the international workshop on reproducibility and replication in recommender systems evaluation (Hong Kong, China) (RepSys '13). Association for Computing Machinery, New York, NY, USA, pp 7–14. https://doi.org/10.1145/2532508.2532511
3. Bottou L, Peters J, Quiñonero-Candela J, Charles DX, Chickering DM, Portugaly E, Ray D, Simard P, Snelson E (2013) Counterfactual reasoning and learning systems. arXiv:1209.2355 [cs.LG]
4. Cañamares R, Castells P (2020). On target item sampling in offline recommender system evaluation. In: Fourteenth ACM conference on recommender systems (Virtual Event, Brazil) (RecSys '20). Association for Computing Machinery, New York, NY, USA, pp 259–268. https://doi.org/10.1145/3383313.3412259
5. Cañamares R, Castells P, Moffat A (2020) Offline evaluation options for recommender systems. Inf Retr J 23. https://doi.org/10.1007/s10791-020-09371-3
6. Chaney AJB, Stewart BM, Engelhardt BE (2018) How algorithmic confounding in recommendation systems increases homogeneity and decreases utility. In: Proceedings of the 12th ACM conference on recommender systems (Vancouver, British Columbia, Canada) (RecSys '18). Association for Computing Machinery, New York, NY, USA, pp 224-232. https://doi.org/10.1145/3240323.3240370
7. Ebesu T, Shen B, Fang Y (2018) Collaborative memory network for recommendation systems. In: The 41st international ACM SIGIR conference on research & development in information retrieval (Ann Arbor, MI, USA) (SIGIR '18). Association for Computing Machinery, New York, NY, USA, pp 515–524. https://doi.org/10.1145/3209978.3209991
8. Felicioni N, Ferrari Dacrema M, Cremonesi P (2021) A methodology for the offline evaluation of recommender systems in a user interface with multiple Carousels. In: Adjunct proceedings of the 29th ACM conference on user modeling, adaptation and personalization (Utrecht, Netherlands) (UMAP '21). Association for Computing Machinery, New York, NY, USA, pp 10–15. https://doi.org/10.1145/3450614.3461680
9. Garcin F, Faltings B, Donatsch O, Alazzawi A, Bruttin C, Huber A (2014) Offline and online evaluation of news recommender systems at swissinfo.ch. In: Proceedings of the 8th ACM conference on recommender systems (Foster City, Silicon Valley, California, USA) (RecSys '14). Association for Computing Machinery, New York, NY, USA, pp 169–176. https://doi.org/10.1145/2645710.2645745

10. Gilotte A, Calauzènes C, Nedelec T, Abraham A, Dollé S (2018) Offline a/b testing for rec-ommender systems. Proceedings of the eleventh ACM international conference on web search and data mining (Feb 2018). https://doi.org/10.1145/3159652.3159687

11. Gomez-Uribe CA, Hunt N (2016) The netflix recommender system: algorithms, value business, and innovation. ACM Trans Manage Inf Syst 6, 4, Article 13 (Dec. 2016):19. https://doi.org/10.1145/2843948

12. Gruson A, Chandar P, Charbuillet C, McInerney J, Hansen S, Tardieu D, Carterette B (2019) Offline evaluation to make decisions about playlistrecommendation algorithms. In: Proceedings of the twelfth ACM international conference on web search and data mining (Melbourne VIC, Australia) (WSDM '19). Association for Computing Machinery, New York, NY, USA, pp 420–428. https://doi.org/10.1145/3289600.3291027

13. He X, Liao L, Zhang H, Nie L, Hu X, Chua T-S (2017) Neural collaborative filtering. In: Proceedings of the 26th international conference on world wide web (Perth, Australia) (WWW '17). In: International world wide web conferences steering committee, Republic and Canton of Geneva, CHE, pp 173–182. https://doi.org/10.1145/3038912.3052569

14. Herlocker J, Konstan J, Terveen L, Lui JC, Riedl T (2004) Evaluating collaborative filtering recommender systems. ACM Trans Inf Syst 22:5–53. https://doi.org/10.1145/963770.963772

15. Hu Y, Koren Y, Volinsky C (2008) Collaborative filtering for implicit feedback datasets. In: Proceedings - IEEE international conference on data mining, ICDM, pp 263–272. https://doi.org/10.1109/ICDM.2008.22

16. Olivier J (2019) Revisiting offline evaluation for implicit-feedback recommender systems. https://doi.org/10.1145/3298689.3347069

17. Jeunen O, Verstrepen K, Goethals B (2018) Fair offline evaluation methodologies for implicit-feedback recommender systems with MNAR data

18. Joachims T, Granka L, Pan B, Hembrooke H, Radlinski F, Gay G (2007) Evaluating the accuracy of implicit feedback from clicks and query reformulations in Web search. ACM Trans Inf Syst 25. https://doi.org/10.1145/1229179.1229181

19. Krichene W, Rendle S (2020) On sampled metrics for item recommendation. Association for Computing Machinery, New York, NY, USA, pp 1748–1757. https://doi.org/10.1145/3394486.3403226

20. Ludewig M, Jannach D (2018) Evaluation of session-based recommendation algorithms. User Model User-Adapt Interact 28:331–390. https://doi.org/10.1007/s11257-018-9209-6

21. McInerney J, Brost B, Chandar P, Mehrotra R, Carterette B (2020) Counterfactual evaluation of slate recommendations with sequential reward interactions. Association for Computing Machinery, New York, NY, USA, pp 1779–1788. https://doi.org/10.1145/3394486.3403229

22. Nedelec T, Roux NL, Perchet V (2019) A comparative study of counterfactual estimators. arXiv:1704.00773 [stat.ML]

23. Pal A, Eksombatchai C, Zhou Y, Zhao B, Rosenberg C, Leskovec J (2020) PinnerSage: multi-modal user embedding framework for recommendations at pinterest. Association for Computing Machinery, New York, NY, USA, pp 2311–2320. https://doi.org/10.1145/3394486.3403280

24. Rendle S (2010) Factorization machines. In: 2010 IEEE international conference on data mining, pp 995–1000. https://doi.org/10.1109/ICDM.2010.127

25. Rendle S, Freudenthaler C, Gantner Z, Schmidt-Thieme L (2009) BPR: Bayesian personalized ranking from implicit feedback. In: Proceedings of the twenty-fifth conference on uncertainty in artificial intelligence (Montreal, Quebec, Canada) (UAI '09). AUAI Press, Arlington, Virginia, USA, pp 452–461

26. Rossetti M, Stella F, Zanker M (2016) Contrasting offline and online results when evaluating recommendation algorithms. In: Proceedings of the 10th ACM conference on recommender systems (Boston, Massachusetts, USA) (RecSys '16). Association for Computing Machinery, New York, NY, USA, pp 31–34. https://doi.org/10.1145/2959100.2959176

27. Yi X, Yang J, Hong L, Cheng DZ, Heldt L, Kumthekar A, Zhao Z, Wei L, Chi E (2019) Sampling-bias-corrected neural modeling for large corpus item recommendations. In: Proceedings of the 13th ACM conference on recommender systems (Copenhagen, Denmark) (RecSys '19). Association for Computing Machinery, New York, NY, USA, pp 269–277. https://doi.org/10.1145/3298689.3346996

28. Zhao Q, Chen J, Chen M, Jain S, Beutel A, Belletti F, Chi EH (2018) Categorical-attributes-based item classification for recommender systems. In: Proceedings of the 12th ACM conference on recommender systems (Vancouver, British Columbia, Canada) (RecSys '18). Association for Computing Machinery, New York, NY, USA, pp 320–328. https://doi.org/10.1145/3240323.3240367

# Attentive Hierarchical Label Sharing for Enhanced Garment and Attribute Classification of Fashion Imagery

**Stefanos-Iordanis Papadopoulos, Christos Koutlis, Manjunath Sudheer, Martina Pugliese, Delphine Rabiller, Symeon Papadopoulos, and Ioannis Kompatsiaris**

**Abstract** Fine-grained information extraction from fashion imagery is a challenging task due to the inherent diversity and complexity of fashion categories and attributes. Additionally, fashion imagery often depict multiple items while fashion items tend to follow hierarchical relations among various object types, categories, and attributes. In this study, we address both issues with a 2-step hierarchical deep learning pipeline consisting of (1) a low granularity object type detection module (upper body, lower body, full-body, footwear) and (2) two classification modules for garment categories and attributes based on the outcome of the first step. For the category and attribute-level classification stages, we examine a hierarchical label sharing (HLS) technique in two settings: (1) single-task learning (STL w/ HLS) and (2) multi-task learning with RNN and visual attention (MTL w/ RNN+VA). Our approach enables progressively focusing on appropriately detailed features for automatically learning the hierarchical relations of fashion and enabling predictions on images with complete outfits. Empirically, STL w/ HLS reached 93.99% top-3 accuracy while MTL w/ RNN+VA reached 97.57% top-5 accuracy for category classification on the Deep-Fashion benchmark, surpassing the current state of the art without requiring landmark or mask annotations nor specialized domain expertise.

S.-I. Papadopoulos (✉) · C. Koutlis · S. Papadopoulos · I. Kompatsiaris
CERTH-ITI, Thessaloniki, Greece
e-mail: stefpapad@iti.gr

C. Koutlis
e-mail: ckoutlis@iti.gr

S. Papadopoulos
e-mail: papadop@iti.gr

I. Kompatsiaris
e-mail: ikom@iti.gr

M. Sudheer · M. Pugliese · D. Rabiller
Mallzee, Edinburgh, UK
e-mail: manjunath@mallzee.com

D. Rabiller
e-mail: delphine@mallzee.com

# 1   Introduction

Fashion, being a primarily visually driven domain, has recently attracted the interest of computer vision researchers for numerous tasks, including attribute recognition, landmark detection, outfit matching, and item retrieval [1]. In this study, we address two central pattern recognition problems on fashion imagery, namely clothing category and attribute classification. Contrasted with other domains, fashion datasets tend to be more diverse and fine-grained—relating to categories, patterns, styles, textile materials, colors, length among many others—rendering the training of deep learning computer vision models a rather challenging endeavour. Additionally, categories and attributes follow hierarchical relationships, meaning that certain attributes apply only to specific categories of certain object types (e.g., tie-front > blouse > upper-body). Finally, fashion imagery often depicts multiple garment items per image increasing the complexity of the problem.

To address the aforementioned problems, we propose a hierarchical two-stage deep learning pipeline that employs "hierarchical label sharing" (HLS). Essentially, HLS shares the predicted label of the previous task to the next; meaning the sharing of the object type labels with the category-level classifier and the predicted category with the attribute classifier. Our two-stage pipeline first identifies low granularity object types (upper body, lower body, full-body, footwear) in fashion images and then classifies the corresponding bounding boxes with regards to category and fine-grained attributes. For the second stage, we examine the performance of HLS in two settings: (1) single-task learning (STL w/ HLS) and (2) multi-task learning (MTL) with a recurrent neural network (RNN) and visual attention (VA). The "MTL w/ RNN+VA" method incorporates HLS in the RNN decoder. We train and evaluate both methods on DeepFashion [2], a widely used fashion dataset and the *Mallzee datasets*—created by Mallzee[1]—that also includes footwear. Combining a Faster R-CNN with an InceptionV2 backbone for object type detection and an EfficientNet-B4 architecture for category and attribute classification, we were able to surpass the state of the art on category classification on the DeepFashion dataset. More specifically, "STL w/ HLS" scored 93.99% top-3 accuracy, while "MTL w/ RNN + VA" scored 97.57% top-5 accuracy; the highest achieved scores for the dataset. A significant advantage of HLS is the ability to learn hierarchical relations among fashion attributes/categories/object types without requiring manually crafted rules by domain experts. Additionally, "MTL w/ RNN+VA" has the advantage of producing attention plots which are useful for interpreting the model's predictions. Finally, our two-stage pipeline has the ability to work efficiently with real-world fashion imagery without requiring further types of annotation.

---

[1] Mallzee is a fashion product aggregator company that provides mobile e-commerce services (https://mallzee.com/).

The main contributions of our work can be summarized as follows:

- We propose two novel hierarchical methods for category and attribute classification on fashion imagery, "STL w/ HLS" and "MTL w/ RNN+VA", that compete with the domain's state of the art.
- We utilize a two-stage pipeline recognizing patterns in full-scale fashion imagery depicting multiple garments.
- We expand our analysis to footwear that are missing from DeepFashion (DF1) [2] and DeepFashion2 (DF2) [3] benchmarks.

## 2  Related Work

Recently, research on fashion-related image classification has received a lot of attention from multiple deep learning disciplines, including image processing [1], multimodal [4, 5], and scene graph learning [6]. In this section, we will mainly focus on the first, image processing, since it is more relevant to our work. Generally, category classification on fashion imagery is formulated as a multi-class classification problem while attribute classification as a multi-label classification problem whose objective is to identify fine-grained attributes relating to styles, patterns, fabric, and length among others.

DeepFashion is a publicly available and widely used fashion-related dataset, created by Liu et al. [2]. In the original paper, Liu et al. (2016), proposed FashionNet in order to assess the usefulness of DeepFashion. FashionNet jointly learned to predict clothing landmarks and attributes. The model was trained end-to-end to first estimate the landmark locations, pool/gate the extracted features and then identify the relevant attributes. Since the publication of DeepFashion and the development of FashionNet, the former has been used as a benchmark dataset and the latter as a baseline model for category and attribute classification in fashion.

Corbiere et al. [7] utilized a weakly supervised approach that learns from large-scale noisy data. Their model was trained contrastively, with the use of negative sampling, from image–text pairs with noisy and mostly unprocessed texts. By fine-tuning a dense layer on the DeepFashion dataset, this approach was able to outperform FashionNet in texture and shape-related attributes but not on the overall evaluation.

Wang et al. [8] proposed the incorporation of domain knowledge to improve attribute classification by developing a fashion grammar capturing kinetic and symmetrical relationships between clothing landmarks. The researchers introduced a bidirectional convolutional recurrent network that leveraged their fashion grammar during the landmark prediction process. After being trained for landmark prediction, two branched fully connected layers were added and trained for category and attribute classification.

Ye et al. [9] combined cost-sensitive learning and over-sampling in order to rectify the problem of class imbalance present in DeepFashion. The authors introduced a weighted loss function that only back-propagated the most informative nodes, therefore focusing on minority classes as the training progresses, and combined it with a semi-supervised Generative Adversarial Network for over-sampling the minority classes by producing synthetic samples.

Finally, Li et al. [10], designed a multi-task model that is trained end-to-end for landmarks, category, and attributes detection. They incorporated two knowledge sharing techniques regarding boundary awareness and structure awareness for transferring relevant information across tasks, yielding 93.01% top-3 accuracy on category classification and 59.83% top-3 recall on attribute classification, reaching the currently highest score on DeepFashion. The results for all aforementioned studies can be seen in Table 7.

A limitation of DeepFashion is that its images are annotated with only one clothing item even if more garments are visible. This phenomenon can create challenges during both training and evaluation stages [3]. To mitigate this problem, all aforementioned research works applied cropping of the training and testing images using the ground truth bounding boxes. However, models solely trained on cropped images can only extract local features from the image and thus can not generalize to images with complete outfits. To overcome this issue, we propose a hierarchical pipeline that separates the tasks of object type detection, category classification, and attribute classification. A second limitation of DeepFashion is its lack of footwear, a significant product category for the fashion industry. For this reason, by utilizing Mallzee's database, we created a new dataset that includes rich annotations about categories and attributes related to footwear in addition to upper body, lower body, and full-body garments.

## 3  Methodology

### 3.1  Problem Formulation

Pattern recognition on fashion imagery is considered a rather challenging task due to the inherent diversity and complexity of fashion items and their relations [1]. Fashion items tend to follow hierarchical relations between different object types, categories and attributes. "Object types" are considered high-level descriptors of garments denoting their relation to the human body. Garments can be classified into upper body, lower body, full-body garments and footwear. Furthermore, fashion items can be classified into various categories and attributes. In this context, category classification is defined as a multi-class task consisting of mid-level descriptors such as "dress", "shirt", "trousers", while attribute classification is defined as a multi-label task with fine-grained descriptors such as "floral", "pencil", "frill". In fashion, object types, categories and attributes tend to follow hierarchical relations. Following the previous examples, a "dress" belongs to the "full-body" object type and "shirt" in "upper-body". The "frill" attribute is applicable to "dresses"—as a specialised "style" of "dress"—but not, for example, to footwear. However, other attributes such as "floral" can theoretically be applied to all categories of garments since it describes the print or pattern of the garment. Another common challenge in the fashion domain is that fashion imagery often depict complete outfits. Thus, in production settings, a pattern recognition model should be able to recognize multiple object types and their location in full-scale images. To address both challenges, we propose a hierarchical two-stage deep learning pipeline that utilizes hierarchical label sharing (HLS) for automatically learning relations among garment object types, categories, and attributes in full-scale fashion imagery.

## 3.2 Proposed Architectures

In this study, we attempt to improve the classification of garment categories and fine-grained attributes by automatically learning the existing hierarchical relations among garment attributes, categories, and object types while being able to perform predictions on fashion imagery with complete outfits. To this end, we propose a hierarchical two-stage deep learning pipeline that employs hierarchical label sharing (HLS). HLS shares the predicted labels from the previous stage to the next; meaning that the object type is shared with the category classifier and the category label is shared with the attribute classifier. Our proposed pipeline follows two stages (1) object type detection and (2) category and attribute classification. A similar method has been applied for self-driving cars [11] but to the best of our knowledge, this is the first time it is attempted in the fashion domain.

### 3.2.1 First Stage: Object Type Detection

The first stage of our deep learning pipeline performs object type detection that identifies the object type and location (bounding boxes) of fashion-related objects on full-scale images. We utilized transfer learning where object type detection models, pre-trained on large-scale image dataset, are fine-tuned on a new dataset. We experimented with variants of Faster-RCNN [12], CenterNet [13], and EfficientNet-D1 and D2 [14]. These models were fine-tuned on the *Mallzee dataset* and the DF2 datasets; discussed in Sect. 4.1. Afterward, we passed the images through the fine-tuned object type detection model and extracted the bounding boxes of each detected garment and its object type label.

### 3.2.2 Second Stage: Category and Attribute Classification

The second stage of our proposed pipeline performs category and attribute classification employing HLS. We examine how HLS works in two settings: (1) single-task learning with hierarchical label sharing (STL w/ HLS) and (2) multi-task learning with RNN and visual attention (MTL w/ RNN+VA); that employs HLS in the RNN decoder. However, in order to assess the effectiveness of HLS we also define a conventional "STL" method that does not utilize HLS (referred to as "Baseline STL").

**Baseline Single task learning (STL).** The training workflow for the Baseline STL follows a conventional image classification process. The images are first passed through the fine-tuned object detection model and the identified fashion-related object types are cropped around the predicted bounding boxes. The cropped images are passed through an image augmentation pre-processing layer that performs alterations to the images as a method of regularization and by extension the mitigation of overfitting. More specifically, the images are horizontally flipped at random and are randomly rotated and zoomed by $\pm 10\%$. The augmented images are then passed through the base convolutional encoder model. The features extracted from the last convolutional layer of the base model are pooled with the use of global average pooling thus transforming the output into a 2D tensor. A classification dense layer is added on top. In the case of category classification (a multi-class problem), the dense layer is activated by a softmax, while for attribute classification (mutli-label task), the last dense layer is activated by a sigmoid function. Subsequently, the model is trained by an adaptive gradient descent optimizer (Adam or RMSProp) which performs gradient updates per individual parameter. The network's loss is calculated by the categorical cross-entropy and the binary cross-entropy for category and attribute classification, respectively. The workflow is shown in Fig. 1a (without the intermittent lines).

**Hierarchical label sharing (HLS).** Our first approach, "STL w/ HLS", follows the same workflow as "Baseline STL" with the addition that the predicted labels from the previous stage are passed to the next. This means that the object type labels for each image are shared with the category classifier, after being passed through an embedding layer and then concatenated with the image representation extracted from the convolutional backbone. Similarly, the garment-level category is shared with the attribute classifier. We hypothesize that HLS will be instrumental in automatically learning hierarchical relations among fashion attributes, categories, and object types. The workflow for this method is shown in Fig. 1a when applying the red intermittent lines.

**Multi-task learning with RNN and visual attention.** Our second approach, "MTL w/ RNN+VA" method relies on multi-task learning (MTL) where the same neural network is fine-tuned by two separate loss functions simultaneously. The network is optimized for both category and attribute classification simultaneously based on the categorical cross entropy and the binary cross-entropy loss functions, respectively. Our MTL architecture—partly inspired by [15]—integrates HLS in a recurrent neural network (RNN) decoder, more precisely a Gated Recurrent Unit (GRU). Moreover, we employ the attention mechanism proposed by [16] in order to enable the model to focus only on the relevant part of the image for the prediction of certain categories and attributes. A notable advantage of this approach is the ability to plot attention weights and thus interpret the model's predictions.

More specifically for our implementation, each image first passes through a vision encoder convolutional neural network (CNN) in order to extract its visual features $F = [f_1, f_2, \ldots, f_n]$, where $f_i \in \mathbb{R}^d$. The attention mechanism receives as input the sequence of vectors $F$ along with the previous GRU hidden state $h_{t-1} \in \mathbb{R}^s$ and

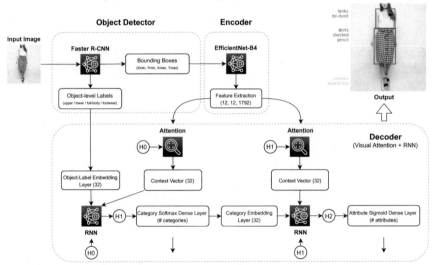

(a) Workflow for "single-task learning" (STL) and "single-task learning with hierarchical label sharing" (STL w/ HLS) methods. The red intermittent line is only applied on the STL w/ HLS method where the label from the previous stage is shared with the next; meaning the object type label is shared with the category classifier and the category label is shared with the attribute classifier.

(b) Workflow for the "multi-task learning with RNN and visual attention" method relying on an encoder-decoder architecture. The vision encoder produces the image representations which are passed through an attention mechanism. In the initial stage, the attentive context vector is given to the RNN alongside the object type label embeddings and the hidden state 0 (H0), a series of zero values. The first resulting hidden state (H1) is passed through two fully connected layers of which the latter is activated by a softmax function with units equal to the number of categories. The context vector is re-calculated given the same image features and the H1 which are passed through the RNN alongside the predicted category embeddings and the H1. The predicted H2 is passed through two fully connected layers of which the later, with units equal to the number of attributes, is activated by a sigmoid function.

**Fig. 1** Workflows for the proposed hierarchical methods

calculates the "context" vector $c_t \in \mathbb{R}^d$ with $t = 1, 2$. The attention weights $a_{ti}$ and the context vector $c_t$ are calculated as below:

$$e_{ti} = V \cdot tanh((W_1 f_i + b_1) + (W_2 h_{t-1} + b_2)) \tag{1}$$

$$a_{ti} = \frac{exp(e_{ti}))}{\sum_{j=1}^{n} exp(e_{tj})} \tag{2}$$

$$c_t = \sum_{j=1}^{n} a_{tj} f_j \tag{3}$$

where $V \in \mathbb{R}^u$, $W_1 \in \mathbb{R}^{u \times d}$, $W_2 \in \mathbb{R}^{u \times s}$ and $b_1, b_2 \in \mathbb{R}^u$ are trainable parameters.

The calculated context vector is then concatenated with the object type embeddings and given as input to the decoder network in order to calculate the next hidden state:

$$h_1 = GRU(h_0, [c_1; E_o]) \tag{4}$$

where $E_o$ are learnable embedding vectors for each object type $o$ and $h_0$ is initialized with zeros. The output of the decoder $h_1$ is then passed through two fully connected layers:

$$U_2^c(U_1^c(h_1)) \tag{5}$$

of which the final $U_2^c$ classifies the image into a garment category, through softmax activation. Consequently, the category and the decoder's previous state $h_1$ are given again to the decoder which performs the same process for the attribute classification:

$$h_2 = GRU(h_1, [c_2; E_c]) \tag{6}$$

$$U_2^a(U_1^a(h_2)) \tag{7}$$

where $E_c$ are learnable embedding vectors for each garment category $c$. The MTL learning pipeline can be seen in Fig. 1b. Finally, the loss function to be minimized is the sum of the two losses (categorical and binary cross entropy) related to category and attribute classification, respectively.

For the second stage, we experiment with InceptionV3, Xception and Efficient-Nets from B0 to B4 with pre-trained weights on ImageNet or from self-training with noisy students [17]. These models consist of relatively limited number of parameters but have been shown to perform very well on the ImageNet benchmark.[2] For each model we also perform hyper-parameter tuning based on the validation accuracy, for identifying the optimal learning rate, dropout rate, batch size, and the optimal number of pre-trained layers to fine-tune.

---

[2] https://paperswithcode.com/sota/image-classification-on-imagenet.

## 4 Experimental Setup

In this section we discuss the datasets and evaluation metrics used in our study. For our experiments, we made use of two publicly available, large-scale, fashion image datasets, DeepFashion (DF1) [2] and DeepFashion2 (DF2) [3]. Additionally, we created a new dataset, which we term *Mallzee dataset*, that also contains annotations for footwear that were lacking from both DF1 and DF2.

### 4.1 Datasets

**Public fashion datasets**. DF1 consists of 289,222 images with rich annotations regarding 46 clothing categories and 1,000 fine-grained attributes related to textures, fabrics, shapes, parts and styles. A significant limitation of DF1 is that each image contains only one annotated garment even if more than one are visible. DF2 consists of 491,895 images in total annotated on 13 categories. Moreover, DF2 expands upon DF1 by including pixel-level mask annotations and by annotating multiple items per image.

We mainly utilize DF1 for the category and attribute classification and DF2 for the task of object type detection. The categories of DF2 are re-mapped as seen in Table 1 to fit our needs for low-granularity fashion-related object types.

Additionally, we calculate the "*Imbalance Ratio*", "*MeanIR*" and "*SCUMBLE*" metrics for each class in order to examine the level of class co-occurrence and imbalance in multi-label data [18].

$$IR(y) = \frac{\arg\max_{y'=Y_1}^{Y_{|Y|}} (\sum_{i=1}^{|D|} h(y', Y_i))}{\sum_{i=1}^{|D|} h(y, Y_i)}, h(y, Y_i) = \begin{cases} 1 & y \in Y_i \\ 0 & y \notin Y_i \end{cases} \tag{8}$$

$$MeanIR = \frac{1}{|Y|} \sum_{y=Y_1}^{Y_{|Y|}} IR(y) \tag{9}$$

$$SCUMBLE(D) = \frac{1}{|D|} \sum_{i=1}^{|D|} [1 - \frac{1}{IR_i} (\prod_{l=1}^{|L|} IR_{il})^{1/|L|}] \tag{10}$$

where D is a multi-label dataset, $Y$ its full set of labels, $y$ the label being analyzed, and $Y_i$ the labelset of the i-th instance in D. $IR$, shown in Eq. 8, is calculated individually for each label as the ratio between the majority class divided by all other classes individually. $MeanIR$, calculated by Eq. 9, simply reflects the mean value across all $IRs$. $SCUMBLE$, calculated by Eq. 10, takes into account both the quotient and product among the $IR$. The initial re-mapped DF2 had a *meanIR* of 1.6528 and a *SCUMBLE* metric of 0.1321 but after randomly down-sampling the dataset, with

**Table 1** Garment categories per object type, for the DeepFashion2 (DF2) dataset and the *Mallzee dataset*. DF2 categories were re-mapped into the object type labels for the object type detection task

| Object type | DF2 categories | Mallzee dataset categories |
|---|---|---|
| Upper-body | Short sleeve top, | Sweaters, Blouses, |
| | Long sleeve top, | Coats and jackets, Formal jackets, |
| | Short sleeve outwear, | Shirts, Hoodies, Camis, |
| | Long sleeve outwear, | Tshirts, Cardigans, Tanks |
| | Vest, Sling | |
| Lower-body | Shorts, Trousers, Skirt | Shorts, Trousers, Skirts, Jeans |
| Full-body | Short sleeve dress, | Dresses, Jumpsuits, Playsuits |
| | Long sleeve dress, | |
| | Vest dress, Sling dress | |
| Footwear | – | Boots, Flats, Heels, Sandals, Trainers |

42,000 objects per class in the training set and 8,000 objects per class in the validation set, the *meanIR* is reduced to 1 and *SCUMBLE* to 0, their optimal values.

*Mallzee dataset*. Considering the importance of footwear in the fashion industry, we deem their lack from both DF1 and DF2 to be a significant shortcoming. Therefore, we decided to create a new dataset that also includes annotated footwear imagery. The dataset is sourced from Mallzee, a fashion e-commerce website, and it comprises three parts, one per task: object type detection, category-level classification and attribute-level classification.

For the object type detection task, we created a dataset with annotated bounding boxes around the garments. The "object type detection dataset" consists of four classes: upper, lower, full-body and footwear. Our objective was to create a balanced dataset in terms of all four classes. To this end, we manually annotated images with bounding boxes with the use of LabelImg.[3] In order to get more images per class, we made use of Haar Cascades[4] for automatically detecting bounding boxes from flat-lay garments. Furthermore, in contrast to DF1—that only contains one annotation per image—we annotated all garments present in an image. During the manual annotation process, our central criterion was that we annotated garments only if they were fully visible; or visible to a significant degree. The resulting "object type detection dataset" was still naturally imbalanced; since for example footwear often come in pairs. To overcome this issue, we created a more balanced dataset by up-sampling the minority classes with the use of image augmentation techniques.[5] More specifically, we applied vertical flips and rotations for flat-lay images and the same plus horizontal flips for manual annotations. The final "object type detection dataset" comprises 5,343 images depicting 15,359 objects: 5,160 upper-body, 3,749 lower-

---

[3] https://github.com/tzutalin/labelImg.

[4] https://github.com/opencv/opencv/tree/master/data/haarcascades.

[5] We did not include instances that contained upper-body garments; being the majority class.

body, 4,985 full-body and 5,554 footwear images. A separate sample of 634 images depicting 1,191 objects (303 upper-body, 205 lower-body, 175 full-body and 508 footwear) is used for evaluation.

For the category-level and attribute-level classification datasets we did not rely on manual annotation. Instead we retrieved the images from Mallzee's database with queries related to 22 garment categories and the 110 attributes classes related to patterns/prints and styles. We queried the target categories and attributes names and all their synonyms we could identify. We applied a series of rules and regular expressions for ensuring minimal mismatch rate.The final datasets comprise 229,633 and 310,753 images for category and attribute classification, respectively. In order to mitigate minor class imbalances we applied image augmentation for the minority classes; similarly to the "object-detection dataset". The 22 category labels can be seen in Table 1 grouped by their object type. Similarly, the attribute key-value pairs can be seen in Table 2, following Mallzee's fashion taxonomy. Table 3, presents and compares the statistics of DeepFashion, DeepFashion2 and Mallzee datasets. The attribute-level Mallzee dataset does not contain all 22 categories. Instead it has 16 slightly broader clustered labels which include: 'activewear', 'boots', 'dresses', 'flats', 'heels', 'jackets', 'jeans', 'shirts', 'shoes', 'shorts', 'skirts', 'sweaters', 'tops', 'trainers', 'trousers', and 't-shirts'. For our experiments, the two datasets were randomly shuffled and split into training, validation and testing sets with 0.8, 0.1, 0.1 ratios respectively.

## 4.2 Evaluation

For the evaluation of the object type detection task we rely on the COCO-challenge metrics[6] for the object type detection metrics, with a focus on the mean average precision (mAP) metric averaged over 10 intersections over union thresholds (IoU) (from 0.5 to 0.95 with steps of 0.05 size)—which is the central metric of the competition—and the average Recall@K (AR@K) that signify the average recall given K detections per image. For category classification we rely on top-K accuracy for K=3,5 and the recall rate at top-K for K=3,5 for attribute classification; since these metrics are also used to benchmark DF1 [2]. For experiments on the Mallzee dataset we also calculate and report the metrics for K=1. All eight methods presented in the results section with which we compare our proposed architectures, have been previously mentioned and elaborated in Sect. 2.

---

[6] https://cocodataset.org/#detection-eval.

**Table 2** Attribute key-value pairs for the *Mallzee dataset* following a taxonomy and the categories they can be applied to

| Att. key | Attribute value | Applies to |
|---|---|---|
| **Prints** | Ethnic, Graphic, Floral, Tropical, | All objects and categories |
| | Striped, Checked, Animal Print, | |
| | Polka dots, Paisley, Spots, Tartan, | |
| | Geometric, Colourblock, Fair isle, | |
| | Camouflage, Grid print, Dip-dyed, | |
| | Tie-dyed, Zigzag, Washed | |
| **Styles** | Duster, Parka, Trench, Peacoat, | Coats and Jackets |
| | Coatigan, Duffle, Bomber, Biker, | |
| | Puffer, Anorak, Windcheater, Borg | |
| | Windbreaker, Coach, Quilted, Trucker | |
| | Wedding, Bridesmaid, Bodycon, Tunic, | Dresses |
| | Jumper, Shirt, Shift, Slip, Tea, Cocktail | |
| | Pinafore, Wrap, Sundress, Smock | |
| | Sweatpants, Leggings, Culottes, Peg, | Trousers |
| | Harem, Capri, Formal, Chino | |
| | Sweatshorts, Cutoff, Bermuda, | Shorts |
| | Skort, Running, Cycling | |
| | Wellies, Winter, Chukka, Chelsea | Boots |
| | Biker, Cowboy, Sock | |
| | Boxy, Varsity, Baseball, Boyfriend, | T-shirts |
| | Fitted, Sport, Polo, Muscle | |
| | Loafers, Brogues, Ballerinas, | Flats |
| | Plimsolls, Boat, Moccasins | |
| | Gladiator, T-bar, Toe-thong | Sandals |
| | Flip-flops, Sliders | |
| | Pencil, Skater, A-line, Frill, Flowy | Skirts and Dresses |
| | Mules, D'orsay, Espadrilles | Flats and Heels |
| | Jeggings, Mom, Boyfriend | Jeans |
| | Tie-front, Bib, Popover | Blouses |
| | Formal, Collarless | Shirts |
| | Waterfall | Cardigans |
| | Blazer | Formal jackets |
| | Cargo | Trousers and Shorts |
| | Court | Heels |

**Table 3** Details for DeepFashion, DeepFashion2 and the *Mallzee dataset*

| Dataset | Images | Categories | Attributes |
| --- | --- | --- | --- |
| DeepFashion | 289,222 | 46 | 1000 |
| DeepFashion2 | 491,895 | 13 | – |
| Mallzee (object-level) | 16,550 | 4 | – |
| Mallzee (category-level) | 229,633 | 22 | – |
| Mallzee (attribute-level) | 310,753 | 16 | 110 |

# 5 Results

## 5.1 Object Type Detection

On the DF2 dataset, CenterNet consistently shows the highest performance; scoring a mean average precision (mAP) of 75.6% and an average recall at 100 (AR@100) of 86.9%. On the other hand, using the *Mallzee* object type detection dataset, a Faster R-CNN model yields the highest scores with a mAP of 80.6% and an AR@100 of 85.8% for 4 classes. The performance of all object type detection models can be seen in Table 4. We use the default hyper-parameters for all models, as given by the TensorFlow object detection API.[7] The only exceptions are the batch size that is reduced to 2 (so as to fit in the GPU memory) while the learning rate is reduced to 1e-4 for EfficientDet-D1 and D2 since the default value is very high and causes overfitting during fine-tuning.

## 5.2 Category and Attribute Classification

### 5.2.1 Results on the *Mallzee Dataset*

For the category and attribute classification tasks we experiment with hierarchical label sharing (HLS) in two settings, (1) single-task learning (STL w/ HLS) and (2) multi-task learning (MTL) with RNN and visual attention (VA). As baseline comparison we define a conventional STL method that does not utilize HLS; named *Baseline STL*. Additionally, we experiment with training four separate STL models, one for each object type. This approach could theoretically improve the specialization of each model and thus improve their predictive accuracy. As can be seen in Table 5, a fine-tuned EfficientNet-B4 for all 22 categories on the *Mallzee dataset*, yields a 92.24% accuracy@1 while the mean accuracy of the four separate models has a slight

---

[7] https://tensorflow-object-detection-api-tutorial.readthedocs.io.

**Table 4** Object type detection models trained on the re-mapped DeepFashion2 (DF2) and the *Mallzee dataset*. Evaluations performed on test sets of each dataset in terms of mean Average Precision (mAP) and Average Recall at 100 (AR@100). The re-mapped DF2 includes three classes (upper/lower/full-body) while the *Mallzee dataset* also includes footwear. (Bold denotes the best performing model by metric)

| Evaluation metric | Training | Evaluation | Faster R-CNN | EfficientDet-D1 | EfficientDet-D2 | CenterNet |
|---|---|---|---|---|---|---|
| mAP | DF2 | DF2 | 73.0 | 72.1 | 64.8 | **75.6** |
| AR@100 | | | 83.7 | 81.9 | 76.7 | **86.9** |
| mAP | DF2 | *Mallzee* | 73.9 | 73.6 | 69.8 | **76.9** |
| AR@100 | | | 84.2 | 82.8 | 80.4 | **86.8** |
| mAP | *Mallzee* | *Mallzee* | **80.6** | 62.6 | 60.5 | 73.2 |
| AR@100 | | | **85.8** | 75.2 | 70.7 | 80.7 |

+0.6% advantage. However, having a single model for all classes will not suffer when encountering misclassified items from the object type detection phase. Meaning that for example, an object wrongly classified as an "upper-body" type while actually being "full-body" will be passed in the wrong model that has not been trained to recognize upper-body type items. We consider that the very slight advantage in accuracy is outweighed by the aforementioned disadvantage. Therefore, having a hierarchical STL architecture—where the output of the object type detection stage is passed directly to separate specialized classifiers—is not deemed optimal.

The results show that the EfficientNet-B4 architecture consistently outperforms all other models when its 100 to last layers (from 474 in total) are fine-tuned with the exception of the batch normalization layers. The aforementioned number of layers concerns the Keras[8] implementation of EfficientNet and includes 'reshaping', 'reduce', 'multiplication' and 'activation' layers as well as the number of convolutional, dense and batch normalization layers. Fine-tuning additional layers does not further improve the model's overall performance while also increasing the required computational resources and time. As expected, we found that low learning rates, either 1e-4 or 5e-5, are optimal for all cases of fine-tuning the pre-trained networks. Moreover, higher dropout rates offer stronger regularization which translates into increased training stability, less overfitting and overall improved performance. Finally, making use of pre-trained weights from self-trained EfficientNets with noisy students [17] improves the accuracy of the architecture when compared to using weights pre-trained exclusively on ImageNet. We apply the aforementioned insights from Baseline STL to both tasks and both methods employing HLS. For the following analysis we use all images of the Mallzee dataset but we did not utilize all 22 categories. Instead we mapped the categories onto the 16 categories found in the attribute-level dataset; so as to ensure the comparability of results.

---

[8] https://keras.io.

**Table 5** Single-task learning models trained and evaluated on the *Mallzee dataset* for category classification. (Bold denotes the best performing model by dataset)

| Training dataset | # of categories | Network | Accuracy@1 |
|---|---|---|---|
| Full-body only | 3 | EfficientNet-B1 | 91.77 |
| | | EfficientNet-B2 | 92.34 |
| | | EfficientNet-B3 | 93.63 |
| | | EfficientNet-B4 | **94.90** |
| | | Xception | 93.46 |
| | | InceptionV3 | 91.55 |
| Lower-body only | 4 | EfficientNet-B1 | 91.79 |
| | | EfficientNet-B2 | 91.46 |
| | | EfficientNet-B3 | 93.16 |
| | | EfficientNet-B4 | **94.44** |
| | | Xception | 93.75 |
| | | InceptionV3 | 92.70 |
| Footwear only | 5 | EfficientNet-B1 | 93.39 |
| | | EfficientNet-B2 | 93.89 |
| | | EfficientNet-B3 | 93.94 |
| | | EfficientNet-B4 | **93.86** |
| | | Xception | 93.17 |
| | | InceptionV3 | 91.87 |
| Upper-body only | 10 | EfficientNet-B1 | 85.31 |
| | | EfficientNet-B2 | 86.04 |
| | | EfficientNet-B3 | 86.33 |
| | | EfficientNet-B4 | **88.15** |
| | | Xception | 86.29 |
| | | InceptionV3 | 83.50 |
| Full dataset | 22 | EfficientNet-B3 | 91.09 |
| | | EfficientNet-B4 | **92.24** |

As illustrated in Table 6, for category classification on the *Mallzee dataset*, STL w/ HLS outperforms the other methods in terms of top-1, top-3 and top-5 accuracy on categories. For attribute classification, HLS does not seem to further improve STL's performance with the exception of a negligible +0.02% in terms of top-5 recall for attributes. MTL w/RNN+VA has the lowest performance of the three settings across all metrics.

Finally, we created a merged pipeline that included the best performing models per task; object type detection, category and attribute classification. The pipeline receives full-scale fashion images which are processed by the object type detector that predicts the garments' bounding boxes. The images are then cropped around their predicted bounding boxes and they are individually passed to the category and attribute classifiers. Three inference examples are presented in Fig. 2.

**Table 6** Comparing the two hierarchical label sharing methods with the baseline STL, for category and attribute classification when trained and evaluated on the *Mallzee dataset*. STL stands for single task learning and MTL for multi-task learning. (Bold denotes the best performance)

| Method | Category | | | Attributes | | |
|---|---|---|---|---|---|---|
| | Top-1 | Top-3 | Top-5 | Top-1 | Top-3 | Top-5 |
| Baseline STL | 90.10 | 98.55 | 99.48 | **78.75** | **94.32** | 96.78 |
| STL w/ HLS | **90.67** | **98.72** | **99.53** | 77.62 | 93.70 | **96.80** |
| MTL w/ RNN+VA | 87.63 | 98.01 | 99.31 | 62.74 | 83.81 | 90.32 |

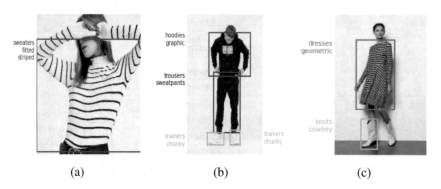

(a)                                (b)                                (c)

**Fig. 2** Inference examples from the *Mallzee dataset* with full-scale images, depicting multiple garments. Red bounding boxes are for upper-body, blue for lower-body, green for full-body garments and yellow for footwear. The images are not part of the training set

### 5.2.2 Results on the DeepFashion Dataset

We trained the three methods (baseline STL, STL w/ HLS and MTL w/ RNN+VA) on the DF1 dataset as a benchmark for category and attribute classification. We did not employ the ground truth bounding boxes offered by DF1 but rather passed its images through our trained object type detection model and only kept the predicted items that matched the images' category level. This decision ensures the generalizability of our method to other datasets that do not include bounding boxes or mask annotations, arguably, two costly and time-consuming types of manual annotation. The object type detector had an 89.2% retention rate meaning that 10.8% of the DF1 was not retained. After randomly sampling 1,000 samples from DF1 we assessed that 3.5% of the mismatch images were due to mistakenly annotated instances in the DF1 while 7.3% were due to mistaken predictions by our model. This translates into approximately 92.7% accuracy for the object type detection model which was not trained on DF1 data.

After cropping the images around the predicted bounding boxes, an EfficientNet-B4 architecture is fine-tuned for category and attribute classification on DF1. The results of each method can be seen in Table 7 compared with all relevant studies. All three models surpass the state-of-the-art on category classification. Specifically,

"STL w/ HLS" has the highest top-3 accuracy with 93.99% (+0.98%) while "MTL w/ RNN+VA" has the highest top-5 accuracy with 97.57% (+0.56%).

On attribute classification, our models do not perform as well, being lower than most previously reported results. This could be an issue resulting directly from our proposed architecture. However, this seems unlikely given the fact that the same network outperforms the state-of-the-art on the category classification task and performs very well on the *Mallzee dataset*. A more plausible explanation is that we are using a slightly different evaluation metric. In the original DeepFashion paper, Liu et al. (2016) used the "recall rate at top-k" metric and reported a 54.61% top-5 accuracy for the attribute classification task [2]. More recently Liu et al. (2020) in MMFashion, reported 14.79% top-5 recall with "VGG + Landmark Pooling" and 30.84% top-5 recall with a "RNN + Landmark Pooling" on attribute classification using the DeepFashion dataset [19]. The employed model and the dataset were the same as the original publication but the end results were vastly different. We could not verify what is the cause of this mismatch. However, it is possible that slightly different evaluation metrics were used. The official DeepFashion GitHub page[9] directs to the MMFashion page[10] whose evaluation protocol[11] uses an altered recall formulation, one that takes only the top-N (N=50) predictions into account while calculating the recall at k. Similarly, we could not verify which evaluation metric was used by all other research teams since their code was not publicly available. In Table 7 we report the conventional recall@k scores. However, we also calculate the aforementioned "altered recall rate" (shown in the parenthesis of Table 7) which results in 65.79% top-3 and 73.57% top-5 by *Baseline STL*, 66.19% top-3 and 73.73% top-5 by *STL w/ HLS*, and 53.01% top-3 and 66.4% top-5 by *MTL w/ RNN+VA*. Again we can observe STL w/ HLS improving upon the performance of Baseline STL for attribute classification. Additionally, if the "altered recall" is indeed the correct evaluation metric for attribute classification on DF1, *STL w/ HLS* has surpassed the current SoTA by 6.36% in terms top-3 recall rate.

Performing an internal comparison, we could see that *STL w/ HLS* improves upon *Baseline STL* on DF1 for both tasks. Especially for attribute classification there are noticeable improvements of 1.49% in top-3 and 1.72% in top-5 recall; using the conventional recall formulation. This is in accordance with our initial hypothesis that HLS can improve image pattern recognition in fashion by capturing existing hierarchical relationships between object types, categories and attributes; without requiring further domain expertise. On the other hand, *MTL w/ RNN+VA* performs very well only on the category-level task but modestly on the attribute-level task. We hypothesize that this disparity can be partly attributed to the fact that the two tasks—while being related to the same fashion images and domain—are relatively dissimilar. By their nature, the task of identifying categories may require the extraction of geometrical features and shapes while attributes may require more

---

[9] https://liuziwei7.github.io/projects/DeepFashion.html.

[10] https://github.com/open-mmlab/mmfashion/.

[11] https://github.com/open-mmlab/mmfashion/blob/150f35454d94a0de7ae40dfdca7193207bd3fc 57/ mmfashion/core/evaluation/attr_predict_eval.py/#L100 .

**Table 7** Benchmarking on DeepFashion for category and attribute classification. In parenthesis we report an "altered recall-rate@k" found in the DeepFashion/MMFashion GitHub page[11] used for the attribute classification task. (Bold denotes the best performance per metric)

| Method | Category | | Attributes | |
|---|---|---|---|---|
| | Top-3 | Top-5 | Top-3 | Top-5 |
| Chen et al. [20] | 43.73 | 66.26 | 27.46 | 35.37 |
| Huang et al. [21] | 59.48 | 79.58 | 42.35 | 51.95 |
| Liu et al. [2] | 82.58 | 90.17 | 45.52 | 54.61 |
| Corbiere et al. [7] | 86.30 | 92.80 | 23.10 | 30.40 |
| Wang et al. [8] | 90.99 | 95.78 | 51.53 | 60.95 |
| Ye et al. [9] | 90.06 | 95.04 | 52.82 | 62.49 |
| Li et al. [10] | 93.01 | 97.01 | 59.83 | **77.91** |
| Liu et al. [19] | – | – | – | 30.84 |
| Baseline STL | 93.71 | 97.40 | 34.71 (65.79) | 43.90 (73.57) |
| STL w/ HLS | **93.99** | 97.49 | 36.20 (**66.19**) | 45.62 (73.73) |
| MTL w/ RNN + VA | 93.72 | **97.57** | 26.85 (53.01) | 35.22 (66.4) |

fine-grained features related to textures, fabrics and styles of the garments. Moreover, in MTL, the same convolutional backbone is being optimized for both multi-class classification with 50 categories and for multi-label classification with 1000 (fairly noisy) attributes. On top of that, the category task reached its peak performance at 10 training epochs and remained stationary while the attribute task required 40 epochs. To this point, a further investigation on alternative ways of calculating and combining the two loss functions (such as weighting) could be insightful for explaining or even mitigating the aforementioned issues.

Despite the limited performance of *MTL w/ RNN+VA*, a noteworthy advantage is the ability to plot the attention weights on top of an image and interpret the model's predictions. Four examples are shown in Fig. 3. Figure 3a depicts a "checked shirt". During the category classification task the model mostly focuses around the shirt's buttons and the neckline area to conclude that the garment is a "shirt". On the other hand, while predicting the garment's attributes, the model broadens its focus and attends multiple points in order to determine that the garment is "checked". In Fig. 3b, depicting a "frill dress", the model gives additional attention to the lower parts of the dress in order to identify its "frill" attribute. On the other hand, Fig. 3c indicates a case where the model performs a correct prediction ("quilted puffer jacket") but the attention plot is not particularly meaningful or interpretable. During the category classification, the model attends more intensely on the shirt under the jacket and not the jacket itself; while correctly classifying it as a jacket. This is a case where the model's prediction is correct but the attention plot is not informative nor interpretable. Finally, Fig. 3d depicts a 'checked shirt', and the model correctly predicts its category ('shirt'). However, the model ignores the hierarchical label sharing information at the attribute stage (that the garment category is a 'shirt') and instead predicts a 'polo';

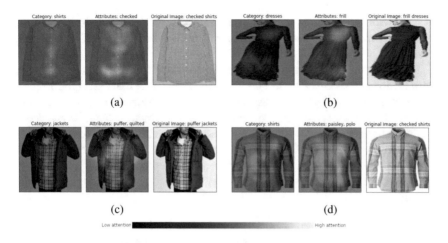

**Fig. 3** Examples of attention plots and their predicted labels from the "MTL w/ RNN+VA" model on images that are not part of the training set. The ground truth labels are reported over the 'original images'

an incompatible combination since the 'polo' attribute only applies to t-shirts. Had the model exhibited higher attention on the long sleeves of the garments it should have taken that into consideration. Furthermore, the model mistakes the 'checked' attribute pattern with 'paisley'; a vastly different type of visual pattern.

## 6 Conclusions

In this study, we propose a deep learning pipeline that employs a hierarchical label sharing (HLS) technique. We examine the performance of HLS in two settings (1) single-task learning with hierarchical label sharing (STL w/ HLS) and (2) multi-task learning with RNN and visual attention (MTL w/ RNN+VA). Our hierarchical pipeline follows two-stages. It first performs object type detection, crops the images around the predicted bounding boxes and then applies category and fine-grained attribute classification. Evaluation on the DeepFashion benchmark shows that our method surpasses the current state of the art (SotA) on category classification. Most notable, "STL w/ HLS" scored 93.99% top-3 accuracy, while "MTL w/ RNN+VA" scored 97.57% top-5 accuracy.

When compared with previous studies, our approach offers the ability to work with full-scale fashion imagery depicting complete outfits. Also, it does not require landmark and mask annotations which are costly and time-consuming types of manual annotation. Furthermore, HLS can learn hierarchical fashion relationships between attributes/categories/types without requiring manually crafted rules by domain experts. Finally, by utilizing the *Mallzee dataset*, the models was shown to cope with footwear, a very significant aspect of the fashion industry, that was completely missing from popular fashion dataset such as DeepFashion and DeepFashion2.

Regarding the further improvement of our models' performance, we consider the introduction of pre-trained human parts detection models in the object type detection phase, in order to improve the precise localization of garments in relation to the human body [22]. Moreover, a hierarchical multi-label loss function could be considered for explicitly discovering meaningful hierarchical relationships among categories and penalizing non-possible multi-label combinations [23]. For future work, we plan on studying how the features extracted from fashion imagery can facilitate the improvement of trend forecasting and garment recommendations in the fashion domain.

**Acknowledgements** This work is partially funded by the project "eTryOn—virtual try-ons of garments enabling novel human fashion interactions" under grant agreement no. 951908. We would also like to thank Jamie Sutherland from Mallzee for his thoughts and input in this work.

# References

1. Cheng W, Song S, Chen C, Hidayati S, Liu J (2021) Fashion Meets Computer Vision: A Survey. ACM Computing Surveys (CSUR). 54:1–41
2. Liu, Z., Luo, P., Qiu, S., Wang, X. & Tang, X. Deepfashion: Powering robust clothes recognition and retrieval with rich annotations. *Proceedings Of The IEEE Conference On Computer Vision And Pattern Recognition.* pp. 1096-1104 (2016)
3. Ge, Y., Zhang, R., Wang, X., Tang, X. & Luo, P. Deepfashion2: A versatile benchmark for detection, pose estimation, segmentation and re-identification of clothing images. *Proceedings Of The IEEE/CVF Conference On Computer Vision And Pattern Recognition.* pp. 5337-5345 (2019)
4. Ma, Y., Yang, X., Liao, L., Cao, Y. & Chua, T. Who, Where, and What to Wear? Extracting Fashion Knowledge from Social Media. *Proceedings Of The 27th ACM International Conference On Multimedia.* pp. 257-265 (2019)
5. Arslan H, Sirts K, Fishel M, Anbarjafari G (2019) Multimodal sequential fashion attribute prediction. Information. 10:308
6. Sadegharmaki, S., Kastner, M. & Satoh, S. FashionGraph: Understanding fashion data using scene graph generation. *2020 25th International Conference On Pattern Recognition (ICPR).* pp. 7923-7929 (2021)
7. Corbiere, C., Ben-Younes, H., Ramé, A. & Ollion, C. Leveraging weakly annotated data for fashion image retrieval and label prediction. *Proceedings Of The IEEE International Conference On Computer Vision Workshops.* pp. 2268-2274 (2017)
8. Wang, W., Xu, Y., Shen, J. & Zhu, S. Attentive fashion grammar network for fashion landmark detection and clothing category classification. *Proceedings Of The IEEE Conference On Computer Vision And Pattern Recognition.* pp. 4271-4280 (2018)
9. Ye, Y., Li, Y., Wu, B., Zhang, W., Duan, L. & Mei, T. Hard-Aware Fashion Attribute Classification. *ArXiv Preprint* ArXiv:1907.10839. (2019)
10. Li, P., Li, Y., Jiang, X. & Zhen, X. Two-stream multi-task network for fashion recognition. *2019 IEEE International Conference On Image Processing (ICIP).* pp. 3038-3042 (2019)
11. Manikandan, N. & Ganesan, K. Deep Learning Based Automatic Video Annotation Tool for Self-Driving Car. *ArXiv Preprint* ArXiv:1904.12618. (2019)

12. Ren S, He K, Girshick R, Sun J (2015) Faster r-cnn: Towards real-time object detection with region proposal networks. Advances In Neural Information Processing Systems. 28:91–99

13. Duan, K., Bai, S., Xie, L., Qi, H., Huang, Q. & Tian, Q. Centernet: Keypoint triplets for object detection. *Proceedings Of The IEEE/CVF International Conference On Computer Vision.* pp. 6569-6578 (2019)

14. Tan, M., Pang, R. & Le, Q. Efficientdet: Scalable and efficient object detection. *Proceedings Of The IEEE/CVF Conference On Computer Vision And Pattern Recognition.* pp. 10781-10790 (2020)

15. Xu, K., Ba, J., Kiros, R., Cho, K., Courville, A., Salakhudinov, R., Zemel, R. & Bengio, Y. Show, attend and tell: Neural image caption generation with visual attention. *International Conference On Machine Learning.* pp. 2048-2057 (2015)

16. Bahdanau, D., Cho, K. & Bengio, Y. Neural machine translation by jointly learning to align and translate. *ArXiv Preprint* ArXiv:1409.0473. (2014)

17. Xie, Q., Luong, M., Hovy, E. & Le, Q. Self-training with noisy student improves imagenet classification. *Proceedings Of The IEEE/CVF Conference On Computer Vision And Pattern Recognition.* pp. 10687-10698 (2020)

18. Charte F, Rivera A, Jesus M, Herrera F (2014) Concurrence among imbalanced labels and its influence on multilabel resampling algorithms. In: International conference on hybrid artificial intelligence systems, pp 110–121

19. Liu X, Li J, Wang J, Liu Z (2020) MMFashion: an open-source toolbox for visual fashion analysis. arXiv:2005.08847

20. Chen H, Gallagher A, Girod B (2012) Describing clothing by semantic attributes. In: European conference on computer vision, pp 609–623

21. Huang J, Feris R, Chen Q, Yan S (2015) Cross-domain image retrieval with a dual attribute-aware ranking network. In: Proceedings of the IEEE international conference on computer vision, pp 1062–1070

22. Gong K, Liang X, Li Y, Chen Y, Yang M, Lin L (2018) Instance-level human parsing via part grouping network. In: Proceedings of the European conference on computer vision (ECCV), pp 770–785

23. Wehrmann J, Cerri R, Barros R (2018) Hierarchical multi-label classification networks. In: International conference on machine learning, pp. 5075–5084

Printed in the United States
by Baker & Taylor Publisher Services